books · music · cafe

books · music · cafe

BORDERS®

Seven Life Lessons of Chaos

Also from HarperCollins by Briggs and Peat:
Turbulent Mirror

Seven Life Lessons of Chaos

TIMELESS WISDOM FROM THE SCIENCE OF CHANGE

JOHN BRIGGS AND
F. DAVID PEAT

HarperCollins*Publishers*

HarperCollins books may be purchased for educational, business, or sales promotional use. For information please write: Special Markets Department, HarperCollins Publishers, Inc., 10 East 53rd Street, New York, NY 10022.

FIRST EDITION

Designed by Jessica Shatan

Briggs, John, 1945–
Seven life lessons of chaos : timeless wisdom from the science of change / John Briggs and F. David Peat. —1st ed.
p. cm.
Includes bibliographical references and index.
ISBN 0-06-018246-6
1. Conduct of life. 2. Chaotic behavior in systems—Miscellanea.
I. Peat, F. David, 1938– . II. Title
BF637.C5B77 1998
158. 1—dc21 97-52983

99 00 01 02 03 ❖/RRD 10 9 8 7 6 5 4 3 2 1

To the memory of David Bohm and David Shainberg,
mariners of the uncertain flow

Lieh Tzu brought a shaman to visit the Taoist master Hu Tzu. But the shaman had trouble making out his face. "Your master Hu Tzu is never the same," complained the shaman. "I have no way to physiognomize him! If he will try to steady himself, then I will come and examine him again."

Lieh Tzu went in and reported this to Hu Tzu.

Hu Tzu said, "Just now I appeared to him as the Great Vastness Where Nothing Wins Out. He probably saw in me the Workings of the Balanced Breaths. Where the swirling waves gather there is an abyss; where the still waters gather there is an abyss; where the running waters gather there is an abyss. The abyss has nine names and I have shown him three. Try bringing him again."

The next day the two came to see Hu Tzu again, but before the shaman had even come to a halt before Hu Tzu, his wits left him and he fled.

"Run after him!" said Hu Tzu, but though Lieh Tzu ran after him, he could not catch up. Returning, he reported to Hu Tzu, "He's vanished! He's disappeared! I couldn't catch up with him."

Hu Tzu said, "Just now I appeared to him as Not Yet Emerged from My Source. I came at him empty, wriggling and turning, not knowing anything about 'who' or 'what,' now dipping and bending, now flowing in waves—that's why he ran away."

But we needn't run.

—Adapted from *The Complete Works of Chuang Tzu*,
translated by Burton Watson

Contents

Acknowledgments

The authors would like to express their gratitude for the help they received from various people along the way to completing this manuscript. Our thanks to Joanna and Maureen for enduring the chaos, Silvio Tavernise, Lucinda Tavernise, Lynda Keen, Michael Patterson, Frank McClushey, Gideon Weil, Kim Witherspoon, and, especially, David Godwin; to our editor, Jeremie Ruby-Strauss, without whom this book would never have come into being and whose idea it was; and to the landscape and people of Pari, Italy, where it was partly written.

Before Words

The Metaphor of Chaos Theory

At one time or another, we've all felt our lives were out of control and heading toward chaos. For us, science has striking news. Our lives are already *in* chaos—and not just occasionally, but all of the time. What's more, the new science suggests, an individual and collective understanding of chaos may dramatically change our lives.

Although we humans tend to abhor chaos and avoid it whenever possible, nature uses chaos in remarkable ways to create new entities, shape events, and hold the Universe together. This revelation about chaos was first made by scientists about thirty years ago and has since been actively investigated. But the real meaning of chaos for us, as individuals and a society, is only now beginning to be explored.

Just what is chaos? The answer has many facets and will take a little explanation. To begin with, chaos turns out to be far subtler than the commonsense idea that it is the messiness of mere chance—the shuffling of a deck of cards, the ball bouncing around in a roulette wheel, or the loose stone clattering down a rocky mountainside. The scientific term "chaos" refers to an underlying interconnectedness that exists in apparently random events. Chaos science focuses on hidden patterns, nuance, the "sensitivity" of things, and the "rules" for how the unpredictable leads to the new. It is an attempt to understand the movements that create thunderstorms, raging rivers, hurricanes, jagged peaks, gnarled coastlines, and complex patterns of all sorts, from river deltas to the nerves and blood vessels in our bodies. Let's begin to grasp this approach by looking at chaos as it is reflected in four very different pictures.

The first photo, taken by the Hubble space telescope, is of a collision between two galaxies. Like a pebble thrown into a lake, this violent encounter flung a violent ripple of energy into space, plowing gas and dust before it at 200,000 miles per hour. This certainly sounds like our traditional idea of chaos, yet within this outer ring of hot gasses, billions of new stars are being born. Here

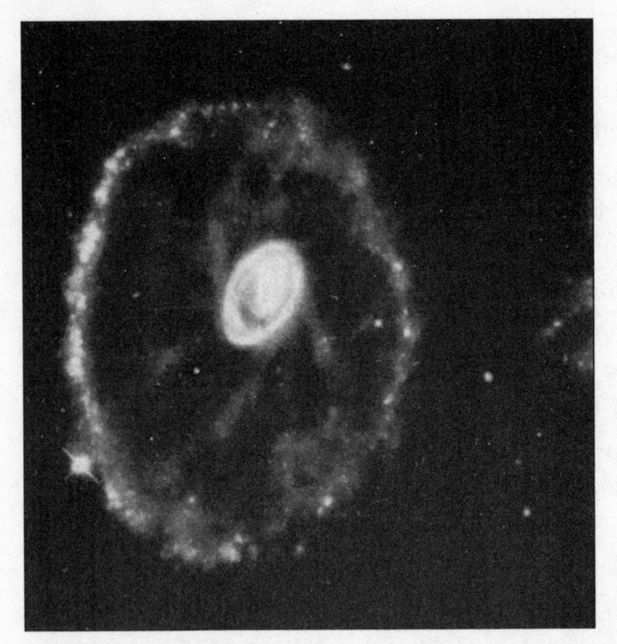

1. *Photo by NASA*

2. *Photo by John Briggs*

3. *Photo by John Briggs*

4. *Generated by Silvio Tavernise*

we see that chaos is both death and birth, destruction and creation. Out of the chaos of primeval gases unfold many varieties of stable order, quite possibly including the highly predictable orbits of planetary systems such as our own. Subatomic particles formed within the first moments of the "big bang" birth of the cosmos are still contained within our bodies in ordered forms. When we die, they will return to the flux of chaos that is as much at work here on Earth as in this galactic explosion. In a deep way, this photograph is a picture of the chaos of each one of us.

The second image shows the turbulence of a mountain stream. Here, apparent disorder masks an underlying pattern. Sit by this stream and you begin to notice that it is simultaneously stable and ever-changing. The water's turbulence generates complex shapes that are constantly renewed. So this stream is another metaphor for ourselves. Like the stream, our physical bodies are constantly being renewed and transformed as cells are regularly replaced. Meanwhile, that "self" that we believe lies within the body at our psychological center is also in flux. We are both the "same" person we were ten years ago and a substantially *new* person. But we can go even further.

A little reflection reveals that the stream depicted here is inextricable from the other ecosystems to which it's connected—the myriad animals and plants that drink from its waters; the twigs, leaves, and seeds that litter the dimple and swirl of its surface; the ancient deposits of glaciers that alter its course; the climate and weather of the region; the season-making orbit of the planet through space. Similarly, each of us as an individual is inter-connected to the systems of nature, society, and thought that surround and flow through us. We live within movements constantly affecting each other and creating an unpredictable chaos at many levels. Yet within this same chaos is born all the physical and psychological order that we know.

The third photograph is an all too familiar image of the everyday human chaos produced by technology and human thought. Vehicles traveling individually along the engineered space of a highway system interact with each other to create alternating regions of gridlocked traffic, sudden stop and go, and free-flowing lanes. Viewed from inside one of those vehicles, the movement of traffic appears patternless and random, but from the perspective of an aircraft flying overhead, subtle patterns emerge—a hidden order within the chaos.

The fourth picture is quite a different image of chaos. Deep within the logically ordered constructs of mathematics lurks a turbulent set of numbers named after Benoit Mandelbrot, the mathematician who discovered them and made them famous. Think of the area depicted within the rectangular frame of the picture as the dense, microscopic rows of dots on a TV screen. Each dot corresponds to a number and is colored as either black or white, depending upon how it reacted when it was slotted into an equation. When the equation was "iterated," or fed back into itself again and again, the number either grew or fell to zero.

The big white warty shape is composed of dots where the numbers fell to zero and stayed there. But in the region along the edge of the white area something strange happens. Here the numbers create patterns that bubble and striate like alien life-forms. The boundaries become filled with all manner of unpredictable repetitions. This bizarre behavior shows that chaos—and its paradoxical order—lies concealed even within the confines of pure mathematical logic. Many people find this mathematical object strikingly beautiful and engaging. Indeed, one of the important characteristics of our new understanding of chaos is its aesthetic appeal.

The scientistic culture that has increasingly surrounded us—and some would say imprisoned us—for the last hundred years sees the world in terms of analysis, quantification, symmetry, and mechanism. Chaos helps free us from these confines. By appreciating chaos, we begin to envision the world as a flux of patterns enlivened with sudden turns, strange mirrors, subtle and surprising relationships, and the continual fascination of the unknown.

Chaos brings us closer to appreciating the world the way artists have appreciated it for thousands of years.

In the past ten years, the idea of chaos has gone far beyond the scientific fields that gave it birth. There are artists who now refer to chaos when talking about their paintings or poems. Chaos theory has figured in hit movies like *Jurassic Park*. The idea is actively being applied to everything from medicine and economics to warfare, social dynamics, and theories about how organizations form and change. Chaos is evolving from a scientific theory into a new cultural metaphor. As a metaphor, chaos allows us to query some of our most cherished assumptions and encourages us to ask fresh questions about reality.

Scientific ideas have blossomed into cultural metaphors previously in history. When Copernicus argued that the Earth revolves around the Sun, his idea did more than just overturn a belief of medieval philosophy; it helped shift Western society's focus from God and the afterlife to individual human beings and the laws of the natural world. As a metaphor, the new image of the heliocentric solar system added great force to the growing European Renaissance and helped the mass of people who were nonscientists and nonphilosophers experience the Universe, and their role in it, in a dramatic new way.

Darwin's theory of evolution had a similar revolutionary effect on the way ordinary individuals viewed the world. The theory showed that we are animals who have evolved within nature. It portrayed life on Earth as the bifurcating branches of a tree. We now view even our own psychology in evolutionary terms. We see ourselves as creatures made up of instincts, drives, and reflexes, as beings who are genetically determined. As a social metaphor, Darwin's notion of "the survival of the fittest" has been used to justify predatory commercial competition and class structure. In fact, the Darwinian idea has become so ingrained that we usually take it for granted that what goes under must have been in some way flawed while what survives must be "better."

What happened to Darwin's scientific idea is an important lesson. Scientific ideas that become cultural metaphors are like medicine. They can be beneficial in the right dosage within the right context, but taken in the wrong way, they can be harmful.

At the moment, it's too early to determine if chaos theory is going to have the same kind of dramatic effect on our consciousness as did these earlier scientific theories. But chaos as a metaphor does have something important in common with them. The idea of chaos opens up radical new ways of thinking and experiencing reality. At the same time, chaos as a metaphor has a built-in humility that previous scientific metaphors did not. Chaos, it turns out, is as much about what we *can't* know as it is about certainty and fact. It's about letting go, accepting limits, and celebrating magic and mystery.

In this book, we will unfold what we see as the metaphor of chaos in the form of 7 lessons, actually 7.1325 . . . lessons (the irrational number 7.1325 . . . is the humility). These lessons are attempts at provocation, penetrations into a new sense of the world, not prescriptions for action or directives about how to think.

Paradoxically, the insights of the newest science share the vision of the world presented in many of the world's oldest indigenous and spiritual traditions. This doesn't mean that chaos theory is about to return us to some mythic golden age or idealized culture, but it does mean that the enduring insights of these cultures will help us elaborate the metaphor of chaos and highlight the way chaos reenvisions ancient wisdom in a brand-new form relevant for our high-tech, high-octane, cyber-saturated age.

Percolating through these lessons of chaos are three underlying themes: Control. Creativity. Subtlety.

First, control. The predicament of all life is uncertainty and contingency. Humans feel this more keenly because our consciousness causes us to remember disasters of the past and imagine dire consequences in the future.

Ancient and indigenous cultures handled their uncertainty through dialogues of ritual with the gods and unseen forces of nature. Western industrial society has taken a different route. We dream of eliminating uncertainty by conquering and controlling nature. The ideal of "being in control" is so much a part of our behavior that it has become an obsession, even an addiction.

Our Western fetish is assailed in Daniel Quinn's novel *Ishmael.* Ishmael satirizes our Western dream of control. We believe, he says, that "only one thing can save us. We have to increase our mastery of the world. All this [environmental] damage has come about through our conquest of the world, but we have to go on conquering it until our rule is absolute. Then, when we're in complete control, everything will be fine. We'll have fusion power. No pollution. We'll turn the rain on and off. We'll grow a bushel of wheat in a square centimeter. We'll turn the oceans into farms. We'll control the weather—no more hurricanes, no more tornadoes, no more droughts, no more untimely frosts. . . . All the life processes of this planet will be where they belong—where the gods meant them to be—in our hands."[1]

Chaos theory demonstrates why such a dream is an illusion. Chaotic systems lie beyond all our attempts to predict, manipulate, and control them. Chaos suggests that instead of resisting life's uncertainties, we should embrace them. This is where the second theme, creativity, enters.

Painters, poets, and musicians have long known that creativity blossoms when they are participating in chaos. Novelists strive for that magical moment when they lose control and their characters take on lives of their own. Old-fashioned logic and linear reasoning clearly have their place, but the creativity inherent in chaos suggests that actually living life requires something more. It requires an aesthetic sense—a feeling for what fits, what is in harmony, what will grow and what will die. Making a pact with chaos gives us the possibility of living not as controllers of nature but as creative participators.

To sacrifice control and live creatively requires attention to the subtle nuances and irregular orders going on around us. Thus, the third theme of this book. The categories and abstractions that constitute our human "knowledge" are certainly necessary for practical survival, but our categories can dominate us to the point where we ignore the finer, uncategorizable inner nature of human situations. We all know that moment when we overreact to something a person has said. We assume that we know exactly what he or she means and we simply can't stand the position they have adopted. In reply, we assert our own opposing point of view and inevitably an argument arises. Chaos suggests an alternative.

Suppose we don't move so quickly to take up a position but instead stay with the original statement and explore the possible inner complexities that lie beyond the other person's abstractions. It could well turn out that the other's abstractions mean something subtly different from what we thought they meant. Or, for that matter, different from what the speaker thought they meant.

The metaphor of chaos theory helps us deal with such situations because it shows that beyond and between our attempts to control and define reality lies the rich, perhaps even infinite, realm of subtlety and ambiguity where real life is lived. Chaos theory shows us how apparently tiny and insignificant things can end up playing a major role in the way things turn out. By paying attention to subtlety, we open ourselves to creative dimensions that make our lives deeper and more harmonious.

In ancient myths throughout history, chaos is central to the creation of the Universe. In Egyptian cosmology, the sun god, Ra, arose from the chaotic waste of flood waters called Nun, while in a Chinese creation story light jumps out of chaos to build the sky. According to the early Greek philosopher Hesiod, "First of all things was *Chaos*."[2]

The clown, trickster, or shape changer becomes the personification of chaos for cultures all over the world. Though he is the "epitome of the principle of disorder," the trickster is also identified as the bringer of culture, the creator of order, a shaman or "super-shaman."[3] The trickster is the wily survivor, the mischievous underdog who defies convention, subverts the system, breaks down the power structure, and gives birth to new ideas. He is the

fox in some traditions, the raven or coyote in others. He is Br'er Rabbit who knows his way around the briar patch. He is Hermes the shape-shifter, Prometheus the fire bringer, Dionysus the god of intoxication and destruction.

Depicted on the ceiling of a cave called Les Trois-Frères in France is a figure from the Ice Age, clearly a shape changer, part man, part animal, perhaps the earliest known recorded trickster. Positioned high in the dome of a subterranean cavern, he overlooks a stunning profusion of beasts and figures painted in momentary turns, leaps, and flashes along the walls. Like a god of chaos who gave birth to these cavorting forms, he confronts us with a wild, penetrating gaze. He is paradox personified—antlered head and ears suggesting a stag, round eyes suggesting an owl, the paws of a bear, the tail of a wolf or wild horse, the prominent sexual organs of a lion, and the beard and chest of a man. His legs are those of a man dancing in what one paleontologist described as a "cake walk." This trickster gazes back at us across thousands of years and bids us join his dance of chaotic transformation at the close of our twentieth century and the beginning of the new millennium.

LESSON 1

Being Creative

Lesson of the Vortex

How did a human being come to make the first arrowhead or the first cave painting? How is it that Einstein discovered the theory of relativity? What happens when we have an original thought? What is the nature of creativity? Why is there something rather than nothing?

Chaos theory offers deep insights into these questions—insights that bear on the nature of each of us as creative beings.

Before exploring these insights, it's helpful to acknowledge that a great many people in our turn-of-the-century society are profoundly ambivalent and misinformed about creativity. If you press them, they'll confess somewhat defensively that they don't really feel they're very creative because creativity is a "gift" or a special "talent" reserved for a few. The idea that true creativity is limited only to special individuals is one of our great myths.

Ironically, although people admire the products that creators make (poems, paintings, music, scientific discoveries), even believ-

Francis Gardner Curtis Fund, courtesy of the Museum of Fine Arts, Boston

ing that creators get to the essence of life, they have at the same time an image of creators as being a little crazy. It's often said that creativity and madness go hand in hand. This fits conveniently with the idea that creativity is somehow abnormal.

Many of us believe that creators exert control over their works (implicit in the high school literature teacher's favorite question, "What did the poet mean to say here?"), while at the same time subscribing to the view that creativity is essentially inspiration over which creators have no control (it just "comes to them").

We also believe that a person can only be creative by working in one of the recognized creative fields like music, film, painting, theater, or higher mathematics. We wouldn't apply the word "creativity" to acts such as observing nature, remembering a dream, talking with someone, or encountering a work of art. Yet poets and other artists have themselves long recognized that such acts are profoundly creative.

A final myth—and one that is difficult to root out—is that the primary goal of creators is to make something new.

The metaphor of chaos theory helps us get beyond these misconceptions and in the process teaches us something "creative" about our lives.

Self-Organization: Nature's Magic

The image that heads this chapter is part of a Chinese scroll painting called "Nine Dragons." In China, dragons are associated with creative power, and here we see a dragon appearing out of a vortex. Like this ancient dragon of chaos, the theory of chaos represents nature in its creativity, embracing a vast range of behaviors, from weather patterns and waterfalls to the firing of neurons and sudden shocks on the stock market. It is as much about how nature makes new forms and structures as it is about nature's "messiness" and unpredictability.

A good example of the broad spectrum of chaotic systems is a river. In the heat of summer, a river runs slowly. Its surface appears calm and serene. Where it encounters a rock, the water parts and flows smoothly past. But in the spring, after heavy rains, the river has a different character. In this circumstance, one part of the river runs slightly faster than a neighboring region and acts to speed up the stream around it, which, in turn, exerts a drag on the faster flow. Each part of the river acts as a perturbing effect on all the other parts. In turn, the effects of these perturbations are constantly being fed back into each other. The result is turbulence, a chaotic motion in which different regions are moving at differing speeds.

As the fast-flowing river approaches the rock, it swirls and turns back on itself. Behind the rock, a vortex is born and persists as a highly stable form. The river is demonstrating all the characteristics of chaos. Its behavior is highly complex, including random, unpredictable flows, eddies, and stable vortices.

The image of the Chinese dragon of creativity coming out of a vortex turns out to be a fortuitous symbol for the theory of chaos. Vortices are superlative—one is tempted to say almost miraculous—examples of the way the zigzags and random traffic of the natural world give birth to structured forms. The vortex of a tornado emerges out of intense thunderstorm activities and turbulent air. The well-known vortex of Jupiter's Red Spot, first noticed in

1664, seems like a permanent feature, but it's actually a vast eddy rolling between giant air streams that travel around the planet in opposite directions.

Complexity theorists refer to the red spot as tornadoes, and other such phenomena as "self-organization out of chaos" or "order for free." To see how this ordering out of chaos comes about, let's examine the formation of vortices in a pan of water.

Turn on the heat beneath the pan and the expected happens. Because hot water is lighter than cold water, water at the bottom of the pan pushes its way upward. Meanwhile, the heavier, cooler water at the top settles down. This rising and falling creates a chaotic competition. Chaos scientists say this system (the heated cylinder of water) is exercising its maximum "degrees of freedom," the maximum range of behaviors available to the system. In short, the water is boiling.

But what are "degrees of freedom"? Think of an orchestra in which each person, if she chooses to, could tune her instrument in an idiosyncratic way and play a different tune in a different key and tempo. The result would be the sonic equivalent of a pan of boiling water—the maximum possible range of behaviors within the orchestra, the largest degree of freedom.

But orchestras and pans of water can also take on a different life. Chaos scientists discovered that if water is heated in just the right conditions below the actual boiling point, a transformation takes place and the water self-orders into a pattern of geometric vortices. For this to happen, first what is called a "bifurcation point" (point of departure) is reached; then the system transforms itself.

To grasp the idea of a bifurcation point, think of a ball in a pinball machine. The ball runs along a straight track until it hits one of the pins. At that instant, it can be flung off to the left or right. The pin is the bifurcation point in the journey of the ball. In the pan, the bifurcation point marks the moment when one of the random fluctuations in the water becomes "amplified" by creating what is called a feedback loop. This loop begins to link with other fluctuations until many interconnected feedback loops create a series of stable hexagonal vortices, or "cells" like a honeycomb, inside the pan.

This linking involves two quite different kinds of feedback. One kind, called negative feedback, damps and regulates activity to keep it within a certain range. A familiar example of a negative feedback loop is the thermostat on an air conditioner. When the temperature rises to a certain point, the thermostat responds by turning on the refrigeration unit. When the temperature drops too low, the thermostat turns the unit off. Negative feedback also operates throughout our bodies. If the sun is hot, we sweat and cool down. When we are too cold, we shiver to produce heat.

A second kind of feedback, called positive feedback, amplifies

A self-organized hexagonal vortex. *Photo by M. O. Velarde, reprinted from* Scientific American, *July 1980, and the* Journal of Non-Equilibrium Thermodynamics, *22(1), 1977*

effects. This happens, for example, when a microphone is placed too close to a speaker system. The microphone detects minute sounds in the room and feeds them to the sound system, where they are amplified and played through the speaker. In turn, the microphone picks up these louder sounds and feeds them back through the system until they quickly become a brain-splitting shriek. Systems like the chaotic river, which are dominated by positive feedback loops, are turbulent and disorderly, but when negative and positive feedback loops couple together, they can create a new dynamic balance—a bifurcation point where chaotic activity suddenly branches off into order.

In the example of the water in the pan, at the bifurcation point, cellular vortices form with hot liquid rising through their centers and colder liquid descending along the outside (a large negative feedback loop). As one vortex butts up against another, stable hexagonal flowing cell walls are created between the descending cascades of falling cooler water.

This self-organized system of heated water creates its structure by giving up some of the degrees of freedom it would have had if it boiled. Think of it as an orchestra whose members decided they'd rather play in concert. They tune their different instruments to concert A and all play in the same key at the same tempo. The result is harmony, order, and a clearly defined musical structure. In a symphony, when each movement comes to an end, the music reorganizes in a new way, with different degrees of freedom being exploited to involve a new key and a new tempo.

Systems that self-organize out of chaos survive only by staying open to a constant flow-through of energy and material. Vortices in rivers and streams typically emerge out of the swirls of turbulence produced downstream from obstructions in a fast, deep current. Each vortex has a definite shape, but is in reality composed of the material flowing through it. In a similar way, we ourselves are composed of the material constantly flowing through us. Our "shape" is created and sustained by the flux of which we are part. We are what we eat, what we breathe, what we experience from our environment.

Many of the structures we see in nature are examples of self-organized chaos. The cupped, hexagonal patterns on the surface of sand dunes, snow fields, and cloud layers result from chaotically organized vortices of warm air rising into the atmosphere, similar to the pan of water. These vortices remain stable as long as the conditions out of which they were created are kept within certain limits.

Watch a flock of birds taking off from the trees and you'll see another type of self-organization in action. The birds jockey frantically, trying to get free of the maelstrom of their fellows and up into the air, wanting to be part of the group, yet at the same time

Self-organized chaos in a cloud layer. *Photo by John Briggs*

trying to avoid collisions with each other. Computer models show that each individual's attempt to keep minimum and maximum distances from others causes flight paths to couple into feedback loops of attraction and repulsion. Positive and negative feedback balance so that the individual birds appear transformed into a single organism. In a similar way, a flock of sandpipers on a beach can turn as a unit faster than individual reaction times would allow.

Random, highly energetic gases in interstellar space self-organize into galaxies and star systems. During the Earth's geological history, self-organization occurred as water ran across the great erosion channels left by melting glaciers. For one reason or another, some paths of water became amplified—more deeply grooved—and linked into one another, eventually forming the vast dendritic patterns of the relatively stable river systems draining the continents.

A hurricane, one of nature's most impressive and massive self-organized forms. *Photo by National Climatic Data Center, NASA*

Some scientists believe that the complex DNA molecule which contains rules that help guide our own unfolding bodies (rules that are themselves subject to the bubbling transformations of chaos) emerged out of a chemical flux in the early days of the Earth, much as the cellular vortices emerge in the pan of water.

So it turns out that chaos is nature's creativity.

Our bodies are pervaded by chaotic, open systems that allow a constantly creative response to a changing environment. For example, our brain self-organizes by changing its subtle connectivity with every act of perception. The list of ways that nature puts the principle of self-organized chaos to use is virtually endless.

People who regularly engage in creative activities usually resonate immediately with the description of how chaos emerges into form, recognizing that they also collaborate with chaos.

A self-organized vortex in deep space. *Photo by National Optical Astronomy Observatories*

Looking at the way professional creators work with chaos gives us a well-documented glimpse into a process that is actually available to each of us—because the simple truth is that we are all creative.

Chaos and Creativity: Truth and the Connection of the Individual to the Indivisible

For the human animal, creativity is about getting beyond what we know, getting to the "truth" of things. That's where chaos comes in.

We're all necessarily conditioned by society. Our conditioning lays out, with apparent certainty, a seemingly complete picture or map of what reality is and how we're supposed to act in it. We're trained to accept and move about in this reality from the moment we emerge from the womb.

Our habits of thought, opinions, and experiences, even the "facts" of the world, are similar to negative feedback loops that go 'round and 'round to keep us in essentially the same familiar place. Such loops of limiting, negative feedback are obviously needed to keep society stable, but they can also be horribly confining if we come to believe that that's all there is to our lives. The danger we all share is of becoming like Pavlov's dogs—our glands reacting every time the bell rings. And society is full of bells.

Often enough, habits of mind, the supposed certainties of our "knowledge" about the world, produce distortions and deceptions about reality. More important, the opinions and facts that constitute our conditioning may end up obscuring a deeper authenticity and "truth" about our individual experience of being in the world.

What do we mean by "truth"? In a culture of postmodern relativism, the word "truth" has become overloaded with unfortunate associations. It's difficult to use in any authentic way. Many people today understandably avoid it, because those in the past who have claimed to possess truth tended to impose it on others, often by violent means. With all the diversity of our modern world, how are we to choose between the truths offered by various religions and cultures? But truth, in the way we mean it here, can't be possessed and imposed on others.

One of the early meanings associated with the idea of the "true" came in the context of the craftsperson who makes a thing straight and balanced. Similarly, a person's life can be "true" in the sense of moving in a straight way, being undistorted, and responding authentically to the present. Here the word "truth" does not mean something absolute (this truth is *the* truth) or relative (you have your truth and I have mine). Truth is, instead, something lived in the moment and expressive of an individual's connection to the whole.

The Indian philosopher J. Krishnamurti described truth this way: "Truth is not a fixed point; it is not static; it cannot be measured by words; it is not a concept, an idea to be achieved."[1] There is no path to truth, he asserted. Truth cannot be arrived at through technique or discipline or logic. It is not something that we agree or disagree about. Truth is what holds us all together, yet each must find it individually out of the terms and conditions of her and his own unique life.

Novelist Joseph Conrad wrote of truth as "the latent feeling of fellowship with all creation . . . the subtle but invincible conviction of solidarity that knits together the loneliness of innumerable hearts."[2] Conrad believed truth can be found in every place in every moment—in small things as well as in grand things.[3] But we're so caught up with looking at the world through our conditioned ideas, opinions, and emotions about its truth that we often don't see right in front of us the sort of truth to which Conrad is referring.

Grasping the truth of the moment was central to the French painter Paul Cézanne. He strove to record on canvas the exact sensations arising in him as he sat in front of his subject. His aim was not to paint his "idea" or conditioned opinion of a landscape or table of fruit, but the exact truth of his moment-by-moment perceptions as they connected him to the life in front of him. He would make small movements of his head as he painted, each new glance acting to shift the entire scene and calling into question what he had previously seen and painted.[4] His paintings are therefore a series of bifurcation points of vision, constituting what has been called "Cézanne's doubt."[5] Cézanne believed that in the fluctuation of these "little sensations," as he called them, lay the truth of his perception.[6] He encourages us to come into contact with the movement of truth that lies in constantly questioning what we see and think about the world.

Truth and chaos are linked. To live with creative doubt means to enter into chaos so as to discover there the truth that "cannot be measured by words."

Making the Vortex 1: Turbulence

The poet John Keats called the entry into chaos an immersion in "doubts and uncertainties." Think of doubts and uncertainties as a way of extending whatever limited degrees of freedom we have come to accept from life. Artists, healers, and those undergoing life changes open up to the uncertainties, accessing degrees of freedom that can spur new self-organization. Going through the death of a loved one, a divorce, or a period of self-doubt is painful, but often those are the very experiences that bring us to a keen sense of the truth beyond words and a new path in life.

The history of the world's religions is full of stories about mystics and sages who spent time in the "wilderness"—either literally or through some "dark night of the soul" and inner chaos. Healing of mind and body in many traditional societies involves a descent into darkness, chaos, and death. Greek healers encouraged "incubation," in which a sick individual was required to sleep and dream. Using ceremonies designed to loosen the grip of the conscious ego, the sick person was encouraged to let go of the familiar structures of his life and enter the dark world of gods and underground forces. By letting go of consensual structures, a creative self-reorganization became possible.

Native Americans use vision quests or the dark, superheated, claustrophobic interior of the sweat lodge to foster psychic self-organization. Traditional psychotherapies make use of the container of the psychoanalytic hour, in which a patient is encouraged to let go, free-associate, and make contact with the chaotic material buried in the subconscious. From out of that primal chaos something true can self-organize.

Creativity simmers in the sweat lodge, in the exploration of uncertainty, in the sacrifice of the familiar. But it need not be heroic or dramatic. Creativity can occur in a conversation when the turbulence of questioning and exchange gives birth to a subtle, new understanding or a true way of expressing something. It can happen when, in looking at a tree, we momentarily dissolve our "knowledge" of trees and see one particular tree's absolute freshness, the unique turns of its branches, its knots and twists, the play of air and light among its leaves. At that moment, we are seeing the truth of the tree. As psychologist Erich Fromm wrote, for the most part "the tree we ordinarily see has no individuality . . . it is only the representative of an abstraction."[7] And so when we encounter the truth we encounter what the Taoist sage Lao Tzu alluded to when he said, "Existence is beyond the power of words to define. Terms may be used but none of them absolute."[8] Seeing the tree beyond abstraction and the seduction of "the known" involves entering, like Cézanne, into doubts and uncertainties and allowing our abstractions and mental constructions to die or be transformed. When this happens, creative insight self-organizes, catching us unaware with the shock or delight of the unexpected truth, essence, or being lurking even within the "objects" of the "ordinary" and "familiar" world around us.

Perhaps it shouldn't be surprising that a high tolerance for ambiguity, ambivalence, and a tendency to think in opposites are characteristics researchers have found common among creative people in many different fields. But professional creators are not born with their heightened tolerance and oppositional tendencies any more than the rest of us. In fact, sometimes it's the reverse. Nevertheless, they come to understand that in order to be creative, they need to give themselves to sensations of "knowing but not knowing," inadequacy, uncertainty, awkwardness, awe, joy, horror, being out of control, and appreciating the nonlinear, metamorphosing features of reality and their own thought processes—the many faces of creative chaos.

Professional creators have many different ways to "cook" themselves in the sweat lodge of chaos. Some dive headlong and helter-skelter into a creative project, flooding their minds with research. Others pile up journal entries full of stray thoughts and streams of consciousness. Others amass information from contradictory or exotic sources, or use other strategies that have the effect of creating "doubt" and increasing their "degrees of freedom."

The physicist Michael Faraday, for example, immersed himself in the nuances of nature, "the beautiful mingling and gradations of color, the delicate perspective, the ravishing effect of light and shade, and the fascinating variety and grace of outline." He sought sensations where "there is no boundary, there is no restraint . . . "[9] Physicist Jules-Henri Poincaré found black coffee helped his creative processes. He compared his thoughts to the hooked atoms imagined by the Greek philosophers. When his mind was in repose, nothing much happened, but when it was charged with energy, the atoms collided and interlocked until they generated new stable forms.

Making the Vortex 2: Bifurcation and Amplification

Because of their willingness, even outright eagerness, to enter a chaotic state, people who engage in creative enterprises have a different attitude about mistakes, chance, and failure than contemporary society.

Creators know that a drip of paint on the canvas, a slip with the chisel on marble, even a mistake in an otherwise well-planned experiment can create a bifurcation point, a moment of truth that amplifies and begins to self-organize the work. This is far different from our usual attitude where mistakes are dismissed as wrong answers, we try to plan accidents out of our enterprises, and failure is an occasion for shame.

Novelist Henry James coined the idea of "the germ" for the point when amplification takes place. A germ is the seed from which the creative thing flowers. Writer E. L. Doctorow said that his novel *Loon Lake* began when he was driving on a country road and noticed a sign. The sound of it seemed to contain something rich, provocative, mysterious, an implicit story. Doctorow said that what starts a story for him "can be a phrase, an image, a sense of rhythm, the most intangible thing. Something just moves you, evokes feelings you don't even understand."[10]

David Whyte, a poet who has worked for years as a consultant and trainer to business organizations, recommends an approach for solving personal or other problems that makes use of the way bifurcations happen. He suggests summoning an image—perhaps from a dream or other source, something that seems powerful to you—and letting it unfold. "The main point is to live with the image or the dream and let it work its magic on us."[11]

The literature of creativity is full of descriptions of that magical moment when the flux of the creator's perception shifts and the chaos begins to self-organize—moments of the *aha!* A completed creative work is a record of the many small and large germs and *aha*'s that leapt into being as the individual pursued the creative activity.

At various times in our lives, we've all experienced germs and *aha*'s like those professional creators talk about—moments of insight when we see or hear something that would be meaningless, nonsensical, or trivial to someone else, but which seem to set in motion a significant change in our perception, to get to the "truth" of our perception, the authenticity of our experience of life. Such insights happen in psychotherapy, for example. They may appear at times of spiritual rebirth, a coming of age, or arrive as a momentary, penetrating realization about the way things are.

Charles Darwin gives us a glimpse into what takes place during those crucial moments when an amplified germ begins to flower into a new understanding. Darwin possessed all the data he needed for his theory of evolution when he returned, as a young man of twenty-six, from his 'round-the-world voyage on H.M.S. *Beagle*. Shortly afterward the naturalist began a notebook where he coalesced his ideas. The key to his discovery was the "tree-of-life" image, familiar to anyone who has taken high school biology. In the image, different organisms are shown branching off from their ancestors like the branches of a tree. During his incubation period, Darwin drew the tree-of-life image in his notebook on three separate occasions. However, it wasn't until he drew it for the third time that he appeared to "see" its deep significance. Just as Cézanne had to shift his head to see the truth of the scene before him, Darwin needed to shift his mental perspective.

We can understand this shift in terms of self-organization. The first tree-of-life image was a bifurcation point, a germ, a small *aha*. The image seemed somehow important and became amplified in Darwin's mind. As he thought about it in relation to various problems of evolution, the image began to couple his thoughts together. By the third time he drew the image, a self-organization was in full swing. The *aha* grew louder. A new context was emerging. Old facts and questions became realigned through the feedback, and more and more could be seen from a new perspective.[12]

Making the Vortex 3: The Open Flow

When we're being creative in our work and daily life, immersing ourselves in chaos, bifurcation sometimes happens. Then a germ-seed concatenates into the flower of an open, flowing creation.

Mihaly Csikszentmihalyi, a psychologist who has studied creativity for many years, gives a description of what this flow of creativity feels like. He interviewed scores of creators, athletes, mountain climbers, religious mystics, and scientists who told him that "flow" is the period in the creative process when self-consciousness disappears, time vanishes or becomes full, and there is total absorption in the activity. There is an intense clarity about the moment and a sense of clear movement, and there is little or no concern for failure.[13]

Moments of flow and exhilaration are the reward for the previous descent into chaos, uncertainty, discomfort, or shock at simply not knowing. The chaos hasn't ended, of course. It's still there, surrounding and feeding the creative activity, like the turbulence fluctuating behind rocks in a river continuously feeding the vortex it has generated.

The idea of the chaotic openness has been associated with self-organized creativity for thousands of years. The first hexagram of the *I Ching* is Ch'ien, "the Creative," the image of the dragon, which the *Ching* identifies with the electrically charged, dynamic arousing forces of a thunderstorm. The commentary on Ch'ien says, "Its energy is represented as unrestricted by any fixed conditions in space and is therefore conceived as motion."[14]

Artists try to keep a sense of flowing openness going within their creative pieces. That's why they use poetic and literary metaphor, irony, and ambiguity—all techniques that plague readers looking for fixed answers, morals, and certainties. Cézanne made landscapes that re-created in viewers the doubt and open questioning he himself experienced as he looked at the scene with small movements of his head. Many artists agonize over the selection of details because they're afraid their choices will reduce or block up the sense of flow. The French poet Paul Valéry expressed his sense of flow by complaining that for an artist no piece is ever really completed, only abandoned. The French painter Marcel Duchamp half joked that one of his major works was now "definitively *un*finished."

The importance of creative openness is reflected in the talking circle of the Blackfoot people. It is the organizational center of their community, the circle where they make their decisions, but they are always careful to leave a gap for the new to enter. This gap represents the open flow always present within their self-organization.

A few years ago, Buddhist monks were creating a sand painting in a public area in Philadelphia. Each day a woman came to watch them at their work. Then, just as they completed the painting, the woman ran into the center and kicked the sand. The organizers were stunned at such an act of desecration by what they took to be a madwoman. The monks, however, welcomed her intervention, for it allowed them to begin again. It was a kick of chaos for another self-order.[15]

The Vortex and the Paradox of Individuality

The idea of openness and the image of the vortex provide a good way to explore one of the most important of the many paradoxes of chaos.

A vortex is a distinct and individual entity, and yet it is indivisible from the river that created it. Beethoven's late string quartets, a Rembrandt self-portrait, a sculpture by Henry Moore, or a sonnet by John Donne are each unmistakably and uniquely made by the hand of their creator, yet at the same time each reveals truths that relate to everyone.

In a vortex, a constantly flowing cell wall separates inside from outside. However, the wall itself is both inside and outside. The same is the case for the membranes in animal and plant cells. The vortex suggests the paradox that the individual is also the universal: Our creative moments—whether it is looking freshly at a tree or coming to a new understanding about our lives—are moments when we are in touch with our own authentic truth, when we experience our unique presence in the world. But, paradoxically, the experience of a unique presence is also often coupled with a sensation of ourselves as indivisible from the whole.

Creative Chaos Means Each of Us

The Romantics pictured the creator as genius and hero, but this first lesson of chaos is that creativity is available to everyone. We can all access an ability to let the ego die for a while and touch the chaotic ground from which forms and orders are constantly bubbling up. Creativity is not just about what takes place in traditionally recognized creative fields. It's what happens in our small and large moments of empathy and transformation, the moments when we contact our authentically individual and therefore universal experience of truth. The British psychologist N. K. Humphrey claims that our greatest use of the human creative intellect is not in art or science but in the day-to-day spontaneous acts by which we hold our society together.[16]

In spite of this, many of us don't feel creative and persistently block the action of creativity in much of our lives. We lose it in our obsessions with control and power; in our fear of mistakes; in the constricted grip of our egos; in our fetish with remaining within comfort zones; in our continuous pursuit of repetitive or merely stimulating pleasure; in our restricting our lives to the containers of what other people think; in our adherence to the apparent safety of closed orders; and in our deep-seated belief that the individual exists in an irreducible opposition to others and the world "outside" of the self.

Chaos theory teaches that when our psychological perspective shifts—through moments of amplification and bifurcation—our degrees of freedom expand and we experience being and truth. We are then creative. And our true self lies there.

The "self," which our postmodern society has enshrined at the center of reality, is essentially a social construction—a collection of categories, names, descriptions, masks, events, and experiences—a complex ever-changing series of abstractions. By entering the chaos of those abstractions, we touch the magical place where the self is also the "not-self," or, if you like, the larger, chaotic self of the world.

Psychiatrist David Shainberg argued that mental illness, which appears chaotic, is actually the reverse. Mental illness occurs when images of the self become rigid and closed, restricting an open creative response to the world.[17, 18]

When water cuts its way through the landscape and self-organizes the sinuous course of a stream, it uses the available materials of rocks, trees, and soil to create its pattern. The key to creative activity lies in the self-organization of available materials. For humans this means we must literally create with our lives. Like water, we can always find a way to be creative with what's available.

Krishnamurti argued that a deep creative appreciation of life comes "only when there is enormous uncertainty."[19] But he saw this uncertainty as existing not just in our grand occasions of life and death but, more important, in each moment. In each moment, we have the opportunity to die psychologically by letting go of prejudices, mechanical habits, isolation, precious ego, images of self and world, and conceptions of the past and future. In this way we set in motion the possibility of a creative, self-organizing perception that puts us in touch with the magic that gave us birth.

Creativity often results in something novel, surprising, and unique, but that's not necessarily its purpose. People usually engage in creative activity because that is where they can contact the authentic truth of the moment in which their individuality converges with something larger. In fact, creativity often involves entering chaos in order to rediscover something old or retrieve the freshness of the everyday. A sense of newness seems an inevitable characteristic of creativity, because when we enter the vital turbulence of life, we realize that, at bottom, everything is always new. Often we have simply failed to notice this fact. When we're being creative, we take notice.

Chaos's lesson of creativity is suggested by the following story: Month after month, year after year, a baker got up early to make bread. One day a customer remarked that over the years the loaves he bought always looked about the same and weighed the same, but the bread always tasted surprisingly warm and fresh. The baker said, "The bread may look the same, but every loaf I make is new because that is where I express my creativity."

Every single morning we also have the choice to be open to the creativity of chaos, open to the world around us, open to the possibility that we can make our lives afresh, like the baker's bread.[20]

LESSON 2

Using Butterfly Power

Lesson of Subtle Influence

Measured against the great forces at play in the world, a butterfly fluttering its wings doesn't seem to possess much power. But an ancient Chinese proverb says that the power of a butterfly's wings can be felt on the other side of the world.

Chaos has shown ways in which this proverb may be literally true. As a metaphor, the chaos idea changes the way we think about power and influence in the world and in our individual lives.

The Secret of the Amplified Small

The scientific insight into the butterfly's power came about through the work of Edward Lorenz, a meteorologist who is considered one of the founders of chaos theory. Lorenz was testing a simple model of weather prediction. The model plugged three kinds of data—wind speed, air pressure, and temperature—into

Photo by John Briggs

three equations that were coupled together in such a way that the results calculated from one equation were fed into the others as raw data and then the process was repeated—in other words, a mathematical feedback loop. In this way, the data of a current weather situation were wound around and around into a simulation of what future weather should look like.

Lorenz had completed a long calculation and needed to double-check his results. Because this was in the days before high-speed computers, he decided to take a shortcut, carrying out the computations to only three decimal places instead of his original six. He knew that by doing this he would be introducing a small error of around 1⁄10 of a percent and expected there would be a similar small degree of difference in his weather prediction.

He was consequently shocked at how little similarity the new weather prediction bore to the previous one that had used the same numbers rounded off to six decimals. Lorenz quickly realized what the culprit was. When the results of each stage of his computation were fed back, or iterated, as raw data for the next, the small initial difference between the two sets of data was quickly magnified by feedback into a large difference. The deductions Lorenz drew from this made him one of the founders of chaos theory.

The coupled equations of Lorenz's weather forecasting model describe what mathematicians call a nonlinear system. It is characteristic of such systems that tiny influences—such as a small error in the initial data—can suddenly blow up in a way that transforms the system. Linear systems, the sort described by conventional science, change smoothly under the application of small influences. Gently depress your car's accelerator, and the car will slowly speed up—small effects produce small changes. On the other hand, you can depress the accelerator so that the passing gear kicks in. Suddenly, you are flung back in your seat as the car surges forward. Linearity has given way to nonlinearity.

Rather than seeing the nonlinearity in his weather model as some sort of defect, Lorenz realized that what was happening in his equations was faithful to what is always occurring in real weather. Because weather is a chaotic system full of iterating feedback, it is nonlinear, which makes it incredibly sensitive to tiny influences. This sensitivity comes from the fact that even small increases in temperature, wind speed, or air pressure cycle through the system and can end up having a major impact. Thus, Lorenz pondered, echoing the Chinese proverb, "Does the flap of a butterfly's wings in Brazil set off a tornado in Texas?"[1]

Let's clarify a little what this chaos adage means.

Weather is the moment-by-moment fluctuation taking place within the self-organized system of the climate. Over very long periods of time, the climate remains the same and on average the weather replays the climatic pattern. But when we look at the climate's pattern in detail, we see that the day-by-day weather is subject to the amplifying, bifurcating, constantly shifting effects of its own iterations. Just as a river produces its own contingencies that lead to turbulence and vortices, the weather creates contingencies that produce its own very changeable behavior.

Modern supercomputers start with a huge volume of information about current weather conditions, iterate it through nonlinear equations, and fairly accurately project what the weather will look like one to three days in the future. But projections beyond that time period, or attempts to describe the weather's fine details within a very small region, become increasingly speculative. One of the countless little butterflies left out of the initial data that have been fed into the computer may be out there asserting itself. In a chaotic system, everything is connected, through negative and positive feedback, to everything else. So somewhere in the real world one of those butterfly loops is pushing a front or changing temperature just a little one way or another. Some knife edge is crossed, the total feedback begins to amplify the small into the large, and suddenly the unpredictable takes place.

After Lorenz had made his discovery, scientists began to see nonlinear "butterfly" effects all around them in complex systems: the few grains of pollen setting off an individual's attack of hay fever, the small trigger of sensations that causes a whole bundle of neurons to fire, the rumor that causes a stock to fall, the fast-spreading grievance that ignites a prison riot. Any one of those internal butterfly loops can become amplified through feedback until it transforms the whole situation.

Humans may continue to dream of the power of prediction and control, but chaos theory tells us that most self-organized systems are laced with countless butterflies of many subtle varieties and colors. In nature, society, and our daily lives, chaos rules through the butterfly's power.

The Power of the Powerless

As an idea, power is an important expression of the deep-seated human desire to have an impact on others and feel connected to them.

In our early hunter-gatherer past, when Homo sapiens banded together in small groups, power probably wasn't much of an issue between people. Each individual could have direct influence on the whole group.

Larger communities and cities made it increasingly difficult for the ordinary person to feel significantly interconnected and to have an impact upon the society as a whole. Societies explored different ways of organizing large numbers of people. The best of these communities tried to balance the need for stability in the collective against the need for freedom and creativity in the individual. Greek city-states are one example of a balance that worked fairly well for the free men in the society, fostering immense creativity in the culture. But the system didn't work well for the society's slaves and for women. Inevitably, imbalances developed between those who had the power of influencing the society and those who didn't. Often enough, individuals who felt insecure and disconnected from their fellows were the ones who sought power.

Anthropologists have discovered that the !Kung people of the Kalahari desert in Africa are acutely aware of the dangers of individuals basing their identities on power. When a !Kung hunter returns home with a particularly rich prize to share, his neighbors denigrate the offering instead of thanking him for it. They explain, "When a young man kills much meat, he comes to think of himself as a chief or big man, and he thinks of the rest of us as his servants or inferiors. We can't accept this, we refuse one who boasts, for someday his pride will make him kill somebody. So we always speak of his meat as worthless. This way we cool his heart and make him gentle." Anthropologist Marvin Harris notes that the !Kung do have leaders who speak out and are listened to with a bit more deference. "But they have no formal authority and can only persuade, never command."[2]

In and of itself, of course, power isn't negative. Human beings need to exercise power to survive in nature, divert streams for irrigation, plow the land, and transport goods. But our human investment in power has gone far beyond these uses. Historically, large societies haven't been as insightful as the !Kung about ensuring that the concept of power doesn't dominate relations within the community. In fact, in modern technological societies, the idea of power has acquired a megaton quality. Long ago, power ceased to be only about the ability to survive in nature or make oneself felt among others. Power became focused on control, imposing our individual will, destroying if necessary. Around the world, history and literature are full of stories of people obsessed with power. The twentieth century has put its own indelible stamp on this idea.

Our modern sense of power has its origins in the industrial revolution and the creation of great machines that generated hitherto undreamed of power. As Matthew Bolton, builder of steam engines, put it in 1776: "I sell here, sir, what all the world desires to have . . . Power." A year later, James Watt wrote, "The velocity, violence, magnitude, and horrible noise of the engine gives universal satisfaction to all beholders, believers or not."[3]

Power of this quality and size may have its place in the factory and on the railway line, but when we attempt to apply the ethos of "power as the only real answer" to the subtle workings of human society, things go badly wrong. Power may be useful for a degree of dominance over some of the forces of nature, but it certainly hasn't worked well for controlling our human natures. In modern and postmodern society, spiritual and humanistic values have declined in the face of the rising central value of power.

Obsessions with power surround us today: the power of money, the power of personality, mind power, computing power, organizational power, political power, the power of love, the power of sex, the power of youth, the power of religion, the power to change our genes or our self-images, firepower, the power relationships between one group and another. Newspapers and television shows incessantly gossip about the lives of the powerful—how they exercise power and whether they are gaining or losing it. We have become inculcated with the idea that if only we had enough power we would be free to do and be what we want. We believe that if we had the power to control the situation, we would feel more secure. The idea of control creates an apparent distinction between the controller and what's controlled.

The truth is our obsession with power may be simply the symptom of our sense of our own powerlessness. All around us vast impersonal organizations and societal forces seem to be shaping our destiny. The spread of voice-mail systems has made it almost impossible to speak to a live human being. Because there seems to be nothing we can do about this, we choke with rage when the system cuts us off at the end of forty-five minutes of pushing buttons in response to a machine.

When we say we feel powerless, we mean that we don't feel powerful enough to fight the corporation, the bureaucracy, the system, the other person's strong personality, or even some wayward "other" lurking within our own psyches. We're outgunned.

Adrift in a world of the powerful, how do we proceed? The usual answer is: Try to get some of that power.

But chaos theory suggests another answer. It says that complex and chaotic systems—which means most of the systems we encounter in nature and in society—cannot accurately be predicted or exclusively controlled. Neither can rigid systems be easily budged. However, there's a loophole. What if we acted through the myriad tiny feedback loops that hold a society together? Chaos tells us that each one of us has an unrecognized but enormous influence on these loops. Chaos suggests that although we may not have power of the controller in the traditional sense, we all possess the "butterfly power" of subtle influence.

What is subtle influence?

In an essay of great importance to many living in Eastern Europe during the late seventies, the Czech writer Václav Havel challenged the traditional response of fighting power with power by proposing the action of something fundamentally different that he called "the power of the powerless." At the time, Havel was unaware that his essay was describing in human social terms the action of Lorenz's chaotic butterfly.

The context for Havel's 1978 tract was the communist regime in Czechoslovakia. Havel knew there was little hope that any of the traditional powers—such as a liberating army or an internal uprising—could transform a post-totalitarian society into one that honored human rights and individual freedoms. So he inquired whether there was another kind of power.

Havel came to realize that power within his country—and, by extension, many of the world's powerful organizations and systems—was not maintained by traditional forms of hierarchical leadership. Rather, it was kept in place by the active collusion of society's least powerful members operating within what he termed an "automatism."

Havel's example of collusion and automatism is the greengrocer who puts up a notice in his shop window, "Workers of the World Unite." The sign has arrived along with the fruit and vegetables from the enterprise headquarters, but the greengrocer doesn't display the sign because he has any real desire to tell the public about its ideal. Havel interprets its real message this way: The greengrocer is telling the world, "I, the greengrocer, live here and I know what I must do. I behave in the manner expected of me. I can be depended upon and am beyond reproach. I am obedient and therefore I have the right to be left in peace."[4]

The sign announces the greengrocer's subordination to the inner dynamics of a system that depends on everyone playing along. His action is one of hundreds of thousands of tiny interconnections that hold the system in place and keep individuals following the party line.

Even though we may believe we live in free and democratic societies, we've all had firsthand experience of the collusion and automatism Havel describes. David Whyte, poet and corporate consultant, tells of an employee in a large corporation (we'll call him George) who attended a staff meeting where the boss asked everyone to rate, on a scale of 1 to 10, the boss's new business plan. Most staff members were aware that the plan was poor, with little chance for success, but it was clear what answer the boss wanted. So most of the staffers said 10. One brave person ventured to say 9½. When George's turn came, he was tempted to give the truthful answer, which would be "close to zero," but he, too, gave in and said 10.[5]

How many times have we all said 10 and put our equivalent of the greengrocer's sign in our window? Or we've stood up against such a system only to find we are overwhelmed by the power of all the powerless others with their own signs in the window. Openly challenge the system and we learn that our most intransigent opponents aren't the presumed power holders but ordinary individuals eager to say 10 and prove they're part of the team. They are, of course, reacting out of their own sense of powerlessness, and perhaps in the hope of getting a little bit of power.

In chaos terms, the systems that operate on collusion and automatism are obviously not creative open systems. Rather, their behavior is dominated by a relatively small number of negative feedback loops. The countless small loops—like the greengrocer's sign—are not an expression of creative degrees of freedom, but

Photo by John Briggs

represent microloops locked together in a way that creates one big obsessive repetitive loop that chaos scientists call a limit cycle.

Limit-cycle systems are those that cut themselves off from the flux of the external world because a great part of their internal energy is devoted to resisting change and perpetuating relatively mechanical patterns of behavior. To survive in such rigid and comparatively closed systems, everyone must resign a little—or often a great deal—of their individuality by blending into the automatism. Those who rise "to the top" in such systems are generally the ones who use empty phrases, those mindless formulas that keep the mechanism of collusion together.

Limit cycles are the systems that make us feel powerless. They are the ones we want to change but can't because they appear to resist all our efforts. These systems are everywhere in society. It may be the sort of system that enables the rich and well-connected to get more benefits from the government than the average voters. It may be the company that keeps losing customers because it doesn't have enough people in shipping and nobody in authority even wants to hear about it. It may even be a closed family system in which guilt-ridden parents repeatedly bail their son out of the spots he gets into because of his alcoholism, their attempts to fix the problem are instead enabling it.

Limit cycles can also operate within an individual's psychology. We all know the sort of person who goes through life repeating the same mistakes. He gets out of one destructive relationship only to plunge into another, all the time protesting that this time it's really going to be different.

Our attempt to control or overpower limit-cycle-dominated systems more often than not end up simply reinforcing them. The interwovenness of feedback loops in chaotic systems suggests that in the end it is always the controller who will be controlled—the would-be overpowerer who will be the one overpowered. Chaos says that treating such systems as if we were separate from them is being blind to the truth.

But if it's true that repetitive, power-obsessed systems are held together by our own collusion within the coupled feedback of the limit cycle, then that implies our influence must be enormous. It suggests our influence could be used in a positive way to bring about a more open, creative environment.

The Power of Subtle Influence

Mike Patterson, a trainer of community organizers for the U.S. Department of Housing and Urban Development, describes a community as a "web of small, seemingly unimportant things—perhaps little courtesies, or favors, looking out for others, a smile or a wave to people on the street, and all the other things people used to do. A nurturing, healthy community is a circle, even a basket, held together by mutual trust, respect, and interdependence. Corporations and similar organizations are pyramids, or triangles, and have clearly defined, even sharp, edges and hierarchies with rigid power relationships."[6]

The Polish anthropologist Bronislaw Malinowskii first pointed out how what he termed "phatic speech"—inquiries about the weather or greetings in the street—creates the general atmosphere that holds society together. The Micmac Indians of eastern Canada and New England agree. They say that an individual's most important work of the day is to walk through the community and exchange gossip. Here the content of the gossip is obviously less important than the being of the person exchanging it. That is where each person's real influence lies.

Subtle influence is what each of us exerts, for good or ill, by the way we are. When we're negative or dishonest, this exerts a subtle influence on others, quite aside from any direct impact our behavior might have. Our attitude and being forms the climate others live in, the atmosphere they breathe. We help supply the nutrients for the soil where others grow. If we're genuinely happy, positive, thoughtful, helpful, and honest, this subtly influences those around us. Everybody knows this when it comes to kids. Kids respond to who you are far more than to what you say. But we're all very deeply and subtly affected by the being of others. Just to take one simple example: Scientists who studied older married couples learned that for each partner, the spouse's mood was more important than even the individual's own state of health. A husband could be in poor health, but if his wife was happy, scientists found it was likely that he would feel happy.

Subtle influence in its negative sense—collusion—holds restrictive limit cycles together, but in its positive sense is vital for keeping open systems renewed and vibrant. The metaphor of chaos gives us a new subtle way of thinking about the difference between beneficial and malignant influence.

The subtlety begins with the fact that butterfly power is, of its nature, unpredictable. We lock into society's feedback loops in so many different ways that it's as difficult to guess the long-term effects of our actions as it would be to predict next month's weather. Perhaps for this reason many of the world's wisdom traditions teach that an action should not only take into account the welfare of others in the future, it should be based on the authenticity of the moment, on being true to oneself, and exercising the values of compassion, love, and basic kindness. Positive butterfly power involves a recognition that each individual is an indivisible aspect of the whole and that each chaotic moment of the present is a mirror of the chaos of the future. Remember that Cézanne and Keats suggested that authentic truth is also rooted in a certain kind of attention to uncertainty and doubt. Positive butterfly power, which is really the power of the open system, comes from there.

In a general way, it's not hard to distinguish negative and positive influences. A negative remark can harden our egos. Negative people appear to be locked into a limit cycle of selfishness, greed, anger, disregard for others, and ruthless ambition.[7] Their lust for power has a mechanical quality about it. But we do need to be careful about our judgments. What at first blush may seem a negative influence could turn out to be positive. There is, for example, an appropriate time to be critical, saying no and setting limits.

When a person's aspirations exceed their present abilities or circumstance, an authentic act would be to point this out to them in a kind but clear manner, no matter how much pain this may cause at the time. If George of our earlier example had said "zero" instead of "ten" he might have sparked anger or disappointment in his boss, yet by saying "ten," he was colluding in the perpetuation of a delusion that could have disastrous consequences for the organization and George's own life.

We feel uplifted by a smile or a kind word, yet the apparently positive only works creatively to keep the system open when done with authenticity and in good faith. An educational movement in North America during the 1970s was based on the principles of operant conditioning. Educational theorists argued that because punishment was out of date, children should be encouraged to learn and become well-behaved through a system of rewards called "positive reinforcement." Teachers' instructional manuals contained a hierarchy of reinforcers that were to be learned by heart: "Great," "You've done good work today," and even, "You are worthy of my love." Teachers were encouraged to practice positive facial gestures in front of a mirror. In short, in the name of positive influence, teachers were being forced to behave in mechanical and inauthentic ways that did not reflect the truth of each individual situation. Of course, many children saw through the system and probably despised their teachers for the ways they were acting. Others became dependent on praise to the extent that a neutral remark by the teacher became equated with punishment.

Each of us is a hidden degree of freedom, an angle of a system's unexpressed creativity. Both from "inside" and "outside" (the words are in quotes because in chaos theory "inside" and "outside" are relative terms), the system is susceptible to the amplifying impact of butterfly power. But who should "take credit" for the exercise of this power? As the following example suggests, butterfly power requires a new attitude toward the meaning of power and influence.

People caught in domestic abuse are trapped in a classic limit-cycle system. Over a period of years, a battered wife telephoned local police for help yet refused to make a formal complaint against her husband. When police suggested she should get out of the marriage, she even found ways to excuse her spouse's behavior. Calling the police and getting them to break up the immediate tension became part of the system she was locked into. Then one night, responding officers sat down with the woman and just listened to her in a nonjudgmental way. After several hours of their attention, questions, and encouragement, she decided to apply for a restraining order and eventually obtained a divorce.

The fact is that it's impossible to know what caused the shift in the woman's perception of her situation. Maybe it was the authentic attention of the police officers, but it could have been any one of a host of other factors that became the last nonlinear straw that broke the limit cycle's back. Thus, positive butterfly power goes hand in hand with a need for basic humility, because we realize that the key to change doesn't so much lie in a single individual's action as in the way many different feedback loops interact.

Havel suggests that within this humble power lies our freedom.

During the period of the communist regime, the people of Czechoslovakia believed they were powerless. Yet, as Havel points out, even in those extreme conditions, individuals found ways to engage in authentic individual creativity. He termed their actions "living in truth." In terms of our chaos metaphor, "living in truth" is the simple (though not always easily achieved) course of opening ourselves up to uncertainty, discovering the edge between our individuality and the universal, and acting from that discovery. This is the real power of the powerless. In our authentic realization of the truth of the moment lies our ability to deeply, if humbly, influence even the rigid systems built on automatism and empty phrases.

As we have seen, rigid systems—limit cycles—depend on everyone sacrificing a little of their creative individuality in order to collude with the system. What would happen if the greengrocer simply took his sign out of the window? Havel realized that this tiny act could cascade through the whole. The grocer's removal of the sign would be "a threat to the system not because of any physical or actual power he had but because his action went beyond itself, because it illuminated its surroundings and, of course, because of the incalculable consequences of that illumination."[8]

In point of fact, many Czech writers and scholars acted like that hypothetical grocer and put aside their "success" in official Czech society to exercise their creative freedom by writing with authenticity about what they really believed. Teachers taught people privately what had been kept from them in state schools. Musicians and artists helped create an independent culture. Workers supported and defended each other within state-sponsored unions. All of them were, in effect, taking the sign out of the window and refusing to collude and support the limit cycle of their oppressive society. As it turned out, such creative individuals ended up having a transformative effect on Czech life. In 1989 and 1990, the collective impact of such individual authentic activities helped self-organize the "velvet revolution" that peacefully liberated Czechoslovakia from the restrictions of a post-totalitarian state.

What if we approached the automatisms and inauthenticity of our own environments in this same spirit? Not in the spirit of confrontation that wants to pit power against power, but in the spirit of engaging our own creativity in the circumstances of the moment? If we do that, we will exert our subtle influence, though we may not always see it, and we may never know how it has contributed to the creation or nurturing of an open system.

Let's extend Havel's example and get a feel for the possibilities of a creative response.

When the greengrocer takes his sign out of the window, he is exerting a subtle influence by showing his acknowledgment of the truth. The butterfly consequences for society of a single, individual act are difficult to predict, but for the greengrocer they could be distinctly unpleasant if he is subjected to surveillance and interrogation for his "antisocial behavior." But he is not just limited to a choice of displaying the sign or not. For example, he could leave the sign in the window but begin to discuss with his friends and patrons why he's afraid to remove it and what the sign really means. Or he might say nothing except to himself, refusing to retreat into convenient cynicism about the sign and instead stopping to face his own facts. In the end, the ultimate choice for the grocer is either to continue to collude and sacrifice his individual creativity or to act in some fashion with an authenticity that lives out the creative truth of his insight.

Once the greengrocer realizes his freedom to exert his subtle influence, he becomes an unpredictable element in an otherwise controlled society; in short, he becomes one of society's tricksters.

Trickster figures show the way creativity can overcome overwhelming odds. Tricksters see beyond the limits of the system and bend the rules. For this reason, tricksters make rigid organizations and governments uneasy. Yet it is just such organizations that need them most. When an organization sees only limited possibilities for growth and change, it is because it is accepting outdated boundaries and contexts that serve only the limited purposes of imposed, coercive limit-cycle power.

The environmentalist Joe Meeker points out that dramatic tragedy, where the hero pits himself against the gods and is destroyed by the process, is valued among cultures with Greco-Roman origins. However, most of the world's other peoples emphasize myths and enacted dramas that focus on comedy. Whereas tragedy is concerned with struggles of power, comedy is about tricksters, ambiguity, and the transformation of roles. Whereas tragedy invariably ends in death, comedy ends in marriage, a continuation of society and fertility brought about through tricking the fates, playing on ambivalence, and the crisscrossing of boundaries and limits.[9] Meeker believes it would be far better if we adopt a more gentle and playful attitude by putting on the mask of comedy.

Faced with a formidable opponent, Oriental martial arts masters use a trickster's creative approach. The idea is not to match strength for strength, but to use an intelligent response to the moment to overturn the opponent. The martial artist yields to the power and force of his opponent and applies a mere butterfly's worth of leverage at the crucial instant to turn a frontal attack back onto itself. The essence lies in an attitude of gentleness and calm in the face of extreme violence.

Butterfly power is what happens at the bifurcation points of evolving systems. Subtle influences affected the direction of the roots here. The overall shape of the root system is the result of countless such subtle influences. *Photo by John Briggs*

A tricksterlike movement in England developed woodland burial sites where people could be laid to rest in a simple shroud and remembered by a tree. The organization began in reaction to the cost of funerals and the wasting of wood to make coffins. But the offshoot is the establishment of what, to all intents and purposes, are fifty-six permanent nature reserves. The trickster innovation was to realize that although environmental protesters can always be removed by force, exhuming a body to make room for development is a difficult and time-consuming legal process. In this way, the organizers joke, you can "die to make a difference" to the environment.

In Montgomery, Alabama, during the 1950s, Rosa Parks served as secretary of the local NAACP and tried to register to vote on several occasions. Parks's story demonstrates how the subtle influence of living in truth can sometimes have dramatic and unimaginable effects. It is a trickster story.

In the South of the 1950s, it was nearly impossible for an African American to vote. Parks had had run-ins with bus drivers and was even evicted from buses because, as she put it, "I didn't want to pay my fare and then go around the back door, because many times, even if you did that, you might not get on the bus at all. They'd probably shut the door, drive off, and leave you standing there." This regular routine of humiliation was a part of the segregationist power structure in which Parks lived.

On December 1, 1955, tired after a long day of work, Parks sat down at the front of a Montgomery bus. When the bus driver ordered her to give up her seat for a white man, she refused. Parks wrote later, "Our mistreatment was just not right, and I was tired of it. I kept thinking about my mother and my grandparents, and how strong they were. I knew there was a possibility of being mistreated, but an opportunity was being given to me to do what I had asked of others."[10]

On that particular day, Parks had no idea that she was starting a revolution. She was living in the truth of the moment, a tired, hardworking human being who deserved the seat no less than the white man who demanded she give it up. As it turned out, Parks's truth cut right through the collusion that held the Montgomery Jim Crow system together. The small influence of her personal protest became quickly and unexpectedly magnified by others. Montgomery citizens were stunned. The city's African Americans began a historic 381-day boycott of the municipal buses. They joined the honorable pantheon of tricksters. They walked to work, formed car pools, and remained peaceful despite the flailing pressures of the white power structure. The movement proceeded on the understanding that if the black citizens wanted to transform the limit cycle of the Jim Crow system, they could not treat the white community as separate from them. Martin Luther King urged that the issue not be allowed to degenerate into one group struggling with another for power. He said at a rally, "We are seeking to improve not the Negro of Montgomery, but the whole of Montgomery."[11] From Rosa Parks's moment of truth, the movement flowered. Whites around the nation began to pay attention and join in revulsion against the injustices of segregated transportation. In 1956 the U.S. Supreme Court ruled such segregation unconstitutional.

Answer to the Cynical Realist

Butterfly power underlines just how deeply influential ordinary individuals can be in society. But it also points to the fundamental humility necessary to exert this influence in a positive way. As with the constant random fluctuations in the heated pan of water, we can never be sure how important our own individual contribution will be. Our action may be lost in the chaos that surrounds us, or it may join with one of those many loops that sustain and replenish an open, creative community. On rare occasions, it may even be taken up and amplified until it transforms the entire community into something new. We can't know the immediate outcome. We may never know if or how or when our influence will have an effect. The best we can do is act with truth, sincerity, and sensitivity, remembering that it is never one person who brings about change

Photo by John Briggs

but the feedback of change within the entire system. As Robert Musil says wittily in *A Man Without Qualities*, "The social sum total of everybody's little everyday efforts, especially when added together, doubtless releases far more energy into the world than do rare heroic feats. This total even makes the single heroic feat look positively minuscule, like a grain of sand on a mountaintop with a megalomaniacal sense of its own importance."[12] Butterfly power results from the fact that, as John Donne put it, "no man is an island." We're all a part of the whole. Every single element in the system influences the direction of all other things in the system.

Butterfly power allows for the impossible. Rosa Parks may have thought it was inconceivable that her small action could be central to changing the long-entrenched Jim Crow system. Nevertheless, her own authentic action provided the trigger that allowed many ordinary people to act in the truth of the moment, transforming the consciousness of an entire nation.[13]

The impossible was something we did naturally as children. Later, we grew up into a more rigid conceptual world where boundaries were absolute and the impossible was locked away in a separate compartment from the practical. But chaos theory reminds us that the real world is constantly in flux and any context can and will change. We may discover tomorrow a way of doing things that is inconceivable today.

So although cynical realists argue that human nature can never change from the greedy, self-centered, hierarchical, power-driven consciousness that has dominated history, chaos theory opens the door on such change. It suggests that consciousness is not confined to what is just taking place privately within our individual heads. Consciousness is an open system like the weather. It is shaped by language, society, and all our daily interactions. Each one of us is an aspect of the collective consciousness of the world, and the contents of that consciousness are constantly being altered by the forces of chaos that each of us expresses. The strategies of human nature are not absolutely fixed. Through chaos, one individual or a small group of individuals can deeply and subtly influence the entire world.

Above: organic litter on the banks of the Amazon River in Colombia, South America. *Below*: a fallen, weathering tree in the Wind River range, Wyoming. Both images record the subtle order and constant unpredictable change at work in a landscape where everything is connected to the movement of everything else. *Photos by John Briggs*

Photo by NASA

Chaos involves branching, crumpling, and shifting, as one part of a "system" interacts with its countless other parts. The mouth of the Betsiboka River in Madagascar in this space shuttle photo (*above*) is reminiscent of a squid or root system. Squid, roots, and this river basin evolved in chaotic environments, and their forms reflect that evolution. *Below*, Oregon's Mt. Hood was folded upward by the crunching together of the Earth's tectonic plates. As the mountain rose, its shape was eroded and created by weather, trees, and other organisms in its environment. Chaotic shapes, which are called *fractals*, have the characteristic that the "parts" of the shape repeat the shape at different scales. For example, the jaggedness of any ridge of Mt. Hood roughly resembles the jaggedness of the whole mountain. If a camera zoomed down for increasingly closer views of any one branch of the Betsiboka River, the images would disclose branches within branches, all the way down to the runoff trickling from the slope of a farmer's field.

Photo by Lawrence Hudetz

Photo by Joe Cantrell

Left: a young powwow dancer. Repeating patterns, slightly varied, appear frequently in the design of indigenous costumes. In this outfit, the natural, unpredictable variety of individual feathers combines with symmetry. Notice how the overall pattern of the costume is reflected at different scales, making it fractal. Patterns in nature are similarly constructed. For example, snowflakes are fractals, joining the sixfold symmetry of crystals with the chaos that affects the way each radial "arm" of the snowflake grows. *Below*: a mathematical fractal that was "grown" by using an equation that constantly fed its results back into itself, producing a symmetrical and yet chaotic shape reminiscent of the boy's headdress.

Fractal image courtesy of Art Matrix–Lightlinks, Ithaca, New York

Photo by Lawrence Hudetz

Left: The fractal unfolding of time produced the jagged pinnacles of the canyon in which the Yellowstone River flows. The self-similar pattern of the canyon records the thousands of years of weather that eroded the landscape. *Below*: The nooks and crannies of time are also illustrated by this cascade of flowers in a wall niche of a Medieval Italian village. The patches on the wall are history in small of the village's daily life. The flowing, fractal shapes of the flowers at the moment the photograph was taken mirror the flowing, turbulent pattern of the wall that has resulted from activity of craftsmen and homeowners over the centuries. Architect Christopher Alexander has said that human places like this one have a timeless "quality without name."

Photo by John Briggs

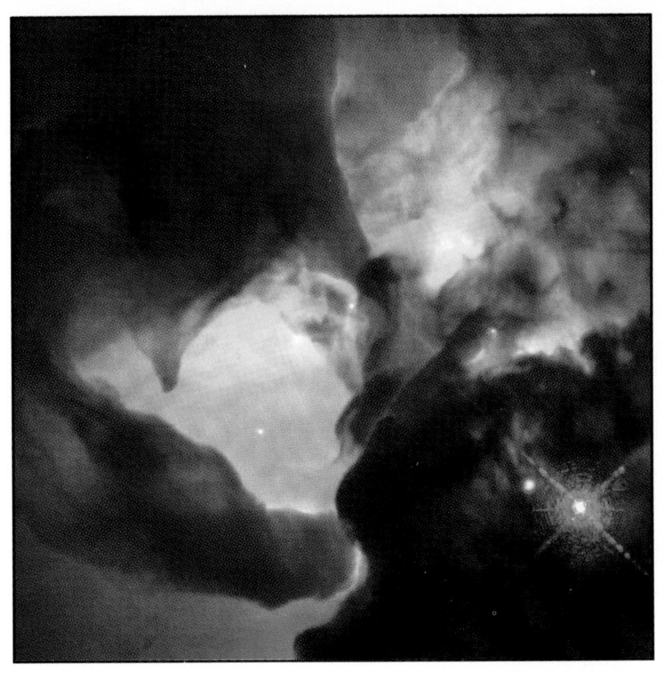

Photo by NASA

The ancient astrologers coined the phrase "As above, so below" to express their belief that the movements of the stars are a mirror of life on Earth. Scientists have discovered that the dynamics that shaped the funnels and boiling rolls of the cosmic-scale Lagoon Nebula (*above*) were at work on Earth creating the much smaller scale erosion structure on the Colorado plateau in Utah (*below*). It's not hard to imagine that these structures also mirror the rolling forms within our own thoughts and emotions.

Photo by John Briggs

Photo by Joe Cantrell

When humans gather together in large groups, they self-organize into fractal patterns—the patterns of chaos. This colorful crowd at a religious festival in Manila resembles the random self-organized order of a field of flowers. Another kind of fractal or chaotic pattern inhabits our logic. Below is a computer-generated "plot" of a portion of the complex number plane. The so-called Mandelbrot set of numbers on the plane behaves in extraordinary, unpredictable ways. When the numbers at the edge of this set are plotted on a computer screen, they show shapes that repeat with endless variations. These patterns, too, are reminiscent of flowers.

Fractal image courtesy of Art Matrix–Lightlinks, Ithaca, New York

Used by permission of Catherine Shainberg

Abstract painting may seem at first merely messy and disordered, but chaos theory suggests that many abstract painters have a deep insight about the patterns of reality. We ordinarily reduce reality to our ideas of trees, stones, animals, and other "objects" and label emotional states with words like *fear*, *joy*, or *tranquility*. *Above*, Painter David Shainberg has stepped back to see the world as a swirl of colors entwined with human emotion. Shainberg titled this painting, with some irony, *A Place of Quiet*. *Below*: Native American Joe Cantrell echoes the painting, capturing one of nature's abstracts in a photograph of lichens he calls *God at Fork Rock* [Oregon]. Cantrell believes that the theory of chaos and fractals has brought Western scientific society closer to understanding nature as native peoples have understood it: not as a collection of objects or as a machine, but as a turbulent dynamical order inseparable from our own perceptions.

Photo by Joe Cantrell

Photo by John Briggs

Landscape artist Nachume Miller doesn't try to paint "realistic" landscapes but hopes to capture the inner dynamics of nature's chaos, the kind of intense activity that takes place in a tumbling waterfall or in the color billowing across a New England autumn. A fragment of one of his pieces is like a single tree in an autumn landscape. Miller says that any fragment will be "very much like the totality of the picture"—each part is a self-similar microcosm of the dynamics of the whole. Miller is one of a group of "fractalist" artists who exploit the ideas of chaos to help them express the way forms—and ourselves—self-organize out of flux.

Oil sketch by Nachume Miller

LESSON 3

Going with the Flow

Lesson about Collective Creativity and Renewal

Wilfred Pelletier, a Native American from an Ojibway community north of Lake Huron, says his people aren't into organization; there's no need for it "because that community is organic." Pelletier gives an illustration of how his unorganized people nevertheless get things done.[1]

"Let's say the council hall in an Indian community needs a new roof. . . . It's been leaking here and there for quite a while and it's getting worse. And people have been talking about it. Nobody organizes a committee or appoints a project leader." Nothing happens, in fact, until "one morning here's a guy up on the roof, tearing off the old shingles, and down on the ground there's several bundles of new, hand-split shakes—probably not enough to do the whole job, but enough to make a good start. Then, after a while, another guy comes along and sees the first guy on the roof. So he comes over and he doesn't say, 'What are you doing up there?' because that's obvious, but he may say, 'How's she look? Pretty rotten, I guess.' Something like that. Then he takes off, and pretty soon he's back with a hammer or a shingle hatchet and maybe some shingle nails or a couple of rolls of tarpaper. By afternoon, there's a whole crew working on that roof, a pile of materials building up down there on the ground, kids taking the old shingles away—taking them home for kindling—dogs barking, women bringing cold lemonade and sandwiches. The whole community is involved and there's a lot of fun and laughter. Maybe the next day another guy arrives with more bundles of shakes. In two or three days that whole job is finished, and they all end up having a big party in the 'new' council hall."

Who was responsible for deciding to put a new roof on the hall? Was it that first guy on the roof, a single isolated individual, or was it the whole community? "How can you tell? No meeting was called, no committees formed, no funds raised. There were no arguments about whether the roof should be covered with aluminum or duroid or tin or shakes and which was the cheapest and which would last the longest and all that. There was no foreman and no one was hired and nobody questioned that guy's right to rip off the old roof. But there must have been some kind of 'organization' going on in all that because the job got done. It got done a lot quicker than if you hired professionals. And it wasn't work; it was fun."

Chaos theory would answer that the "organization" in Pelletier's roofing project was self-organization. It began with chaos—all that disorganized talk beforehand about the leak. The first guy on the roof was a bifurcation point that became amplified. The feedback between the first fellow and the next one who came along started a cascade that coupled the community together around the project, and then the system got the job done.

Clearly, Pelletier's Ojibway community is an open, creative, chaotic, nonlinear system. As he put it, the people in this group "aren't into competition. But they aren't into cooperation either—never heard of either of those words. What they do just happens, just flows along." Within the community's creative open system, micro self-organized systems spring up from time to time, such as the community's action to repair the roof. Such short-term self-organization renews the community and keeps it alive, as testified to by the big party held in the new council hall.

Social self-organization and collective creativity doesn't only happen in Native American communities, it happens it rural communities around the world and in informal organizations of all kinds. In many different circumstances, people start coming together, helping out, lending a hand, throwing in their two cents. Nobody's leading particularly, but things get done.

A high-tech example of social self-organization is the Internet. The Net was started back in the 1960s by the U.S. military looking for a distributed command system in the event of nuclear war so that no single center could be knocked out. The idea was similar to the one that conceived of the U.S. highway system as a distributed airport of landing and takeoff strips. It occurred to the planners that computers all over the country could be linked together to create a giant system that shared its information. But once the Net was set up, academic scientists began to use it and it was eventually made available to the public all over the world. Relatively quickly, more and more individuals and groups joined, until by the mid-1990s an estimated 25 million people were on-line and the number was doubling every eighteen months.

Nobody's controlling the Net (at least not yet). It's maintained by an open flow of users passing information around. Within the global self-organization of the Net and its subset, the World Wide Web, are countless mini self-organizations springing up all the time. People come together to do creative work—everything from photographers displaying their pictures of lightning strikes to underground musicians converging on Web sites to create an audience for their work to interest groups discussing the Vietnam War or Brazilian cuisine. For those who have access, the Net is a daily example of collective creative exuberance. Most of the activity is carried out by people who are making things, looking for information, and exchanging ideas that simply interest them as part of who they are. The giant, hierarchically structured, power-driven commercial organizations have so far been largely frustrated in their efforts to harness the Net to their mechanical engines of profit. Anyone who has surfed the Net knows he has entered a chaotic, dynamic open system where "what they do just happens, just flows along." Clearly there's order here, but it's chaotic.

Taken together, the traditional Ojibway community and the new cyber community suggest a radically different approach to social organization than the one currently taken by postindustrial society.

Life, Complexity, and the Strange Attractor

From the chaos perspective, all activity in society and nature is a collective activity. In chaos, individuals are an indivisible part of the whole. Chaos offers many insights into the curious, paradoxical relationship between individual and group. Take termites, for instance.

When termites gather into a critical mass of numbers they behave differently than they do as isolated individuals. Isolated pairs of termites will mate and lay eggs, but they don't touch each other with their antennae. The mitochondria in their flight muscles are not active. When they come together in a group, however, individual termites undergo chemical and behavioral changes. They begin to touch each other repeatedly with their antennae and show increased metabolic activity in the mitochondria of their flight muscles. In the collective mode, they make nests by engaging in nonlinear activity. First, individuals in the group roam around randomly, carrying and dropping particles of earth. As they wander, they impregnate their earthen dollops with a chemical that attracts other termites. Eventually, by chance, a higher concentration of impregnated earth forms in one area, initiating a bifurcation point. More termites are attracted to the area and their activity couples into the erection of a pillar for their nest. Afterward, termites clean and repair their nest by other kinds of chaotic self-organization.

Children at play regularly self-organize into spontaneous games. *Photos by John Briggs*

Collective Rules and Individual Rules

Within the termite world, individual, semi-isolated termites exhibit one kind of creative behavior (such as mating) while group-level termites exhibit quite a different kind of creative behavior (such as building and cleaning nests). Obviously, building nests requires individuals and mating doesn't make sense unless there's a nest and a collective to take care of the eggs. So, as we might expect with self-organized chaos, we're not talking about any absolute division between individual and group behavior. Nevertheless, there's a clear difference.

Self-organized systems composed of individuals, like the termites, contain varying levels of complexity. Each level has evolved its own "rules." Individual and paired termite behavior follows one set of rules, collective behavior follows another. An important thing to notice is that when a group of individuals comes together, it's not because any single individual or elite is taking the lead. Rather, organization arises from a coupling of feedback springing from random individual activity.

Individuals self-organizing into a protest rally. *Photo by John Briggs*

Such coupling together wouldn't be possible if nature were simply a collection of relatively isolated mechanical parts—the picture that science has given us for the past two hundred years. Through the window of chaos we can now see that the proclivity for individuals to interact and self-organize must be deeply inherent in nature.

Collective Creative Structures

Chaos shows that when diverse individuals self-organize, they are able to create highly adaptable and resilient forms. One good example is the food distribution system of New York City. John Holland, a complexity theorist, noticed some amazing things about this system. Manhattan is an island with no more than a week's supply of food on hand. The system that feeds the city has to respond to the kaleidoscopic transformations the island undergoes every day: There are new buildings being erected and torn down, changes in fads for different cuisines, ever shifting populations. Yet, Holland notes, New York City is free of famines or gluts, and you can find whatever food you want at any time of day or night. The food system percolates efficiently within the fertile boundary between order and chaos.

Holland argues that most of the formal rules (traffic, health and safety, consumer protection, and so on) that help keep things moving along weren't planned in advance, but emerged as the system itself emerged. New York's food distribution system evolved, as chaotic self-organized open systems do, from the bottom up, out of feedback among interacting individual elements. These include individual entrepreneurs, varied groups of consumers, large commercial organizations, and functions of government. Picture what would happen if New York City's government or some privatized entity tried to impose a food distribution system from the top down, setting five-year plans, strategic goals, budgets, forecasts, procedure manuals, and job descriptions.[2] It was just this kind of attempt to "manage" the natural chaos of society with a global plan that the Chinese communists attempted in the 1950s by imposing a command economy. The result was catastrophic shortages and famines.

The Coevolution Perspective

At this point, ardent free-market capitalists may hope to leap on this food distribution example as proof of their view that the best way for individuals to organize and relate to each other is through unfettered competition. However, chaos views the example from its own perspective. According to chaos, believing New York's food distribution system is an entity essentially created by competition is like believing that apples exist because of insecticide.

Capitalist ideology is very similar to the ideology of traditional Darwinian biology, and capitalism has frequently used biology as a rationalization for unrestrained competitive practices. Darwin proposed competition as the major force in the evolution of life, the main energy driving the relationship between individual and group and one group and another.

Chaos theory shifts perspective and allows us to appreciate the fact that biology is full of "coevolution" and "cooperation." These activities probably have a far more significant impact on the shape of things than does competition. As biologist Brian Goodwin puts it, "I'm not denying natural selection. I'm saying that it does not explain the origins of biological form, of the pervasive order we see out there."[3] Rhesus monkeys illustrate Goodwin's point.

According to the theory of natural selection, competition, hierarchy, and dominance (power) are the key survival and reproductive strategies for a species, and therefore for the individuals within that species. Rhesus monkeys have long been considered by biologists one of the quintessential hierarchical primates. Native to India, the rhesus live in troops of about forty, and both males and females in the troops appear to have a clear pattern of ranking. For example, the "alpha" female in the troop can displace any of her underlings at a water fountain, a spot in the shade, or a scramble for food. Similarly, dominance is exerted all the way down the line. In light of Darwinian theory, biologists naturally concluded that the central social activity within the rhesus troops must be an endless competition for dominance.

Darwin asserted that the purpose of the dominance struggle, the big payoff, was that the tougher, "fitter" animals—the ones in the top ranks of the hierarchy—would get to mate more frequently and pass on their genes.

Studying the rhesus using DNA fingerprinting, scientists have discovered, however, that there is something wrong with this competitive picture. Analysis revealed that the high or low rank of a female rhesus bore no relation to her ability to bequeath her own genes, mate with any of the males of the troop, or bring new males into the group. In rhesus society, the females as a group decide what males are "fit" to be allowed into the troop, and if no female shows an interest in a male, it doesn't matter how big he is or how long his teeth are (how dominant he is, in other words), the females can gang up and chase him away.

On the male side, the dominance hierarchy seems to have little to do with how often or with whom a male mates. The important thing appears to be convincing some female you should be a troop member, usually through grooming. Says Kim Wallace, a primate biologist, "The model we have of low-ranking animals striving to be high-ranking animals probably really isn't accurate. The low-ranking animals may be perfectly happy as long as they're getting mating opportunities and as long as they're getting fed."[4]

In fact, the situation where the dominant animals are controlling the genetic destiny of the society would be an abnormal and unhealthy situation. Breeding high aggression and combativeness would risk destroying all the subtle, cooperative behaviors that hold a monkey troop together and ensure the survival of the whole society and its individuals over the long haul.

So letting some nasty, pushy individual full of bellicose juices get ahead of you at the water fountain may not be a sign of weakness but a sign of your strength in knowing the best way to maintain social harmony. Research shows that those individuals bound and determined to "win" at the water fountain are often high-strung, edgy, stress-prone characters not very good at reproducing, and not particularly well adapted. The monkey study indicates that the cooperative, less dominant types are the fittest members of society, if reproductive success is used as a measure.

However, hewing to their understanding of Darwin's ideas, biologists and nature-film makers have focused our attention not only on dominance behavior within a species but on the competitive predator-prey relationships between species. The result is that we have come to think of nature in the stereotype of a one-rule game, "red in tooth and claw." But what about the myriad ways that different species engage in collective creativity through coupling together feedback?

Focusing on this aspect, chaos scientists have found natural history to be filled with examples of what they call "coevolution." For example, 100 million years ago, nature evolved flowering plants with seeds enclosed in fruits, but at the same time animals who enjoyed eating the fruits had to evolve with them. The animals spreading the plants' seeds fostered further experimentation, leading to new plant and animal species. The plant and animal evolution were coupled together in one system.

The rain forest is a delicate and intricate example of coevolution and cooperation. Everything from the fungi that feed on fallen trees to wildly plumed birds and leaf-cutter ants evolved in relation to each other and constantly collaborate with each other in incalculably subtle ways that ensure their mutual survival.

From the perspective of chaos theory, it is less important to notice how systems are in competition with each other than it is to notice how systems are nested within each other and inextricably linked. Competition is a reductionist and limited idea that doesn't begin to appreciate the deep creativity at work in nature.

Competition has become a mental cliché often used to describe behavior that isn't really competitive, reinforcing our belief that the central fact of life is competition. Are the people on the Internet competing with each other? Some are, but most aren't. It seems pretty clear that competition is not what is essentially driving that system, holding it together, making it vibrant.

Of course, competition can be an important element in the way individuals interact. Athletes love the spirit of competition and become exhilarated when pitted against each other. But we should note that their competition takes place within a context of cooperation. Agreements by individuals to cooperate in teams and follow rules make competition possible. Sports fans cooperate by paying admission and cheering and singing together.

Beyond this, one of the most exciting sports experiences anyone can have is watching a team catch fire. Perhaps as a basketball game begins, the players of one team seem to be operating independently of one another, mechanically going through their routines, in effect competing among themselves. Then they suddenly undergo a transformation. One of them makes an inspired play that leads to a basket: At this instant a bifurcation point becomes amplified. Now the moves the players make seem coupled together, all five team members working like a single organism. In this creative self-organization we observe something more than just the competition between two teams.

Chaos theory tells us competition and cooperation are not either/or ideas. They are complexly interwoven.[5]

A complex chaotic system like a rain forest or the human body contains a constantly unfolding creative dynamic in which what we call competition may suddenly become cooperation and vice versa. In chaotic systems, interconnections flow among individual elements on many different scales. In the body, these scales include molecules moving between cells, the cells themselves, tissues, organs, and distributed systems like the immune system and the endocrine system with its hormone secretions from various glands. Instead of seeing these scales of order in terms of competition, chaos focuses on how elements within systems and the relationships between systems are continually reassembling themselves on the edge of chaos.

The "Strangeness" of Chaotic Collectivity

The activity of a collective chaotic system, composed of interacting feedback among its many scales of "parts," is sometimes referred to by the poetic name "strange attractor." When scientists say that a system has an "attractor," they mean that if they plot the system's changes, or behavior, in mathematical space, the plot shows that the system is repeating a pattern. The system is "attracted" to that pattern of behavior, scientists say. In other words, if they perturb the system by knocking it away from the behavior, it tends to return to it fairly quickly.

In the restrictive limit-cycle system, behavior is mechanically repetitive, with fewer degrees of freedom. The system goes through its restrictive behavior independent of what is going on outside. The pattern of a strange-attractor system, however, is different. The strange-attractor plot shows that the system's behavior is unpredictable and nonmechanical. Because the system is open to its external environment, it is capable of many nuances of movement.

For the heart muscle, the attracting behavior is a firing sequence of neurons. The heartbeat rhythm we're all familiar with is produced by this sequence. Scientists studying the sequence discovered that it contains something "strange." The behaviors of mechanical systems such as pistons and clock pendulums are consistent and regular. Their behavior can be mathematically plotted as smooth circles or other shapes. Not so with the heart. Even though we think of the heart as relatively mechanical in its beats (one reason we refer to the heart as a "ticker"), the fact is a healthy heart *isn't* quite regular. It exhibits a strangeness that involves endless chaotic variations, microjolts, and tiny fluctuations within each heartbeat. When the heart's behavior is plotted, the attractor shape folds around itself, this way and that, revealing this strangeness.[6]

Tiny fluctuations in cardiac rhythm are, in fact, a sign of the heart's health, a display of its robustness. The neurons firing in sequence to contract the cardiac muscle don't behave anything like

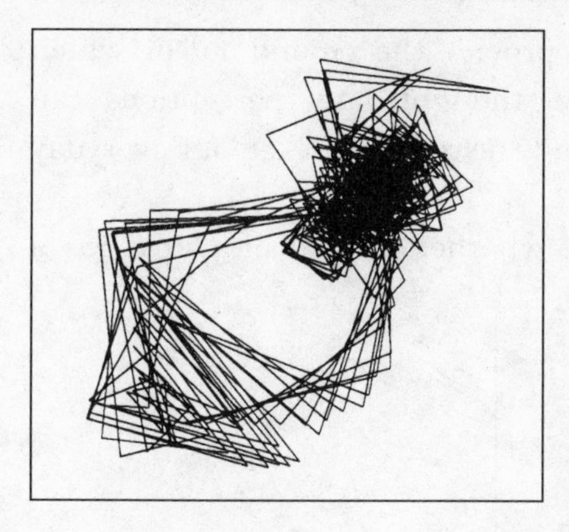

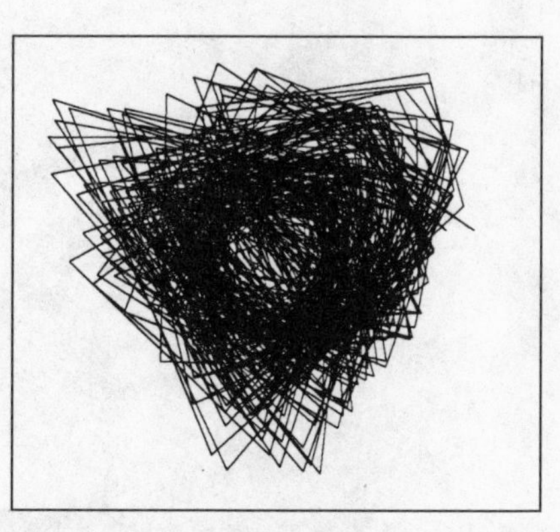

These are plots of heart rhythms. The first shows the strange attractor behavior underlying a normal cardiac rhythm. The second plots the beat pattern of a diseased heart, showing that the rhythm is more mechanical and less chaotic. The patient died of a heart attack eight days after this measurement was taken. *Plots by Dr. Ary Goldberger*

a series of spark plugs mechanically igniting within an engine cycle. Instead, they're a self-organized chaos. This chaos gives the heart a range of behavior (degrees of freedom) that allows it to settle back into its rhythm even after it has been nudged away by some shock such as a fast run or a sudden step into subzero air. Physicians have even discovered that if they detect a heartbeat becoming increasingly mechanical and regular, that's a signal of problems, a sign the heart lacks flexibility. It's become brittle. Now if it's nudged slightly, it won't return to its rhythm, but may simply stop altogether or go wildly careening off into the so-called bag of worms firing called defibrillation.

The attractor for the brain is even stranger, requiring a constant high level of neuronal chaos to provide the ground out of which the sudden self-organization of thoughts and perceptions can arise. Chaos, it turns out, is behind the scenes of even our everyday experience of reality.

Overall, a healthy organism, whether animal or plant, *has* a

These two photos were taken at different times from the same angle. They show that the flow in this streambed, while unpredictable from moment to moment, maintained an overall strange attractor shape. A vortex is a strange attractor with fewer degrees of freedom than are exhibited at this place in the stream. *Photos by John Briggs*

strange attractor and *is* a strange attractor—jiggling, moving, shifting, filled with positive feedback loops that push the system into new directions and negative feedback loops that keep processes from flying off into merely random oblivion. Within the overall strange attractor of the organism lurk many subsets of strange attractors (for example, the heart and the brain), each with its own particular degree of "regularity"—each more or less strange, in other words.

Diversity and Open, Chaotic Systems

One of the vital principles of strange attractors and collective chaos involves the sheer *diversity* of all these systems within systems. A healthy ecology contains a wide range and variety of species interacting with each other. If we reduce the variety and make the system more homogenized, it becomes brittle and is liable to collapse nonlinearly.

Chaotic creativity suggests why diversity is so important. When diverse individuals come together, they have a tremendous creative potential. For example, according to biologist Lynn Margulis, on the early Earth, oxygen-breathing bacteria in search of food invaded other bacterial cells. The invader and host began to develop feedback that allowed the host cell to breath oxygen and gave the invader a supportive environment. This symbiosis led to the kinds of cells we have in our own bodies.

As individuals—each with their own self-organized creativity—couple together, some degrees of freedom are given up but other degrees are discovered. A new collective intelligence emerges, an open system, unpredictable from anything one could have expected by observing the individuals acting in isolation.

Our Persistent Illusion

On the surface, at least, the way we human individuals organize ourselves in modern society doesn't look very much like the chaos view of self-organized forms discussed here. Most of our formal organizations, with their hierarchical organizational charts, mission statements, and annual reports don't resemble Wilfred Pelletier's Ojibway community, the organization of the Internet, or New York City's food distribution system.

The structures we work in and that govern our society are derived from a markedly different set of assumptions about reality. In fact, those assumptions have created our reality, or more accurately the illusion we take for reality. It's a reality in which we worship power and believe that having it is an essential for survival. It's a reality in which we see the world in terms of winners and losers, where we submit to hierarchical pecking orders and tacitly acknowledge the ideology that those at the top are somehow better than those who aren't. It's a reality where we form ourselves into groups and social organs that resist diversity and where our social structures operate as closed entities, many deriving their identity from their opposition to other groups.

Our governments, the corporations we work for, even the leisure and religious groups we belong to sometimes do terrible things in our name. When that happens we blame the leaders or the others in the group. We feel detached from the collective activity that we are an integral and colluding part of. At one level, we may identify totally with an organization, although at another, we feel the organization is an alien other, a *them.* The chaos perspective allows us to see that our distress has a lot to do with how we have bought into the assumptions that organizations are essentially maintained by leadership, competition, and power.

These assumptions pervade our society so thoroughly that they are for the most part invisible. As often happens with invisible beliefs, they infiltrate our observations about the world so that the world seems again and again to confirm them. One of the science writers who reported on the controversial new discoveries about rhesus monkeys noted that "sometimes the ones who are the most obsessed with determining the dominance ranking" were not the monkeys but "the scientists doing the observing."

The toll of our dominance view of reality is insidious and often frightful. In his book *In the Absence of the Sacred*, Jerry Mander shows how TV—which moment by moment reasserts the values and logic of winners and losers, better and best, heroes and leaders—has exerted a corrosive influence on Native American youth. Thanks to its influence, the Native American values of cooperation and sharing are being replaced by competition and rivalry.[7]

In the broader culture, the logic of our assumptions has contributed to a dehumanizing process: a belief that the power of mechanisms, plans, and technologies can save us; the creation of widespread societal passivity and despair; a monoculture riven by ethnic and racial strife; a culture chained to schedules and accomplishments ("having it all") to the extent that individuals seem to have less and less time for simply being; a culture obsessively fascinated with celebrities, images, charisma, and upward mobility.

In 1909 the German sociologist Max Weber warned, "It is horrible to think that the world could one day be filled with nothing but those little cogs, little men clinging to little jobs and striving for bigger ones. . . . It is as if . . . we were deliberately to become men who need to order, who become nervous and cowardly if for one moment this order wavers, and helpless if they are torn away from their total incorporation in it."[8]

Scholars have calculated that individuals in so-called hunter-gatherer or subsistence societies needed only about eighteen hours a week to provide for their food and shelter. In modern postindustrial societies, most of us spend sixty to seventy hours a week at "work" and much of the rest of the time recovering from the stress. People spend far more time at their jobs than they do with their families or in the spiritual contemplation of the mysteries of life. Even the most "primitive" subsistence societies seemed to find plenty of time for those renewing activities.

For most of us, life revolves around our jobs. But in our jobs we are less and less nourished and the organizations we work in are increasingly mechanical and impoverished. David Whyte describes himself as a poet who has tried to bring "soul" into American corporations. He says, "The overripe hierarchies of the world, from corporations to nation-states, are in trouble and are calling, however reluctantly, on their people for more creativity, commitment, and innovation." But this call comes at the same time that the closed, hierarchical, competitively organized, and linearly planned structure of organizations are hell-bent on preventing those creative qualities from ever self-organizing within corporate walls.[9]

People give energy to their work as if to a sacred activity that would make them whole and alive, but too often the job leaves individuals alienated, hyperactive, divided, and depressed. Whyte points out that, at work, people are not allowed to admit weakness, acknowledge self-doubt, or make mistakes without paying heavily for it. In other words, expressions of chaos are suppressed—the very activities necessary for creativity to take place. Whyte also points out that, deprived of the vital fluid of creativity from its members, "the hierarchical systems based on power emanating from the top cannot plan for the wild efflorescence of impossible events we call daily life."[10] Echoing management guru W. Edwards Deming and others, Whyte believes that an organization willing to honor the soul of its workers would be a stronger, more ethical, and less destructive structure. It "would be an organization willing to ask deeply radical questions about whether its products are actually necessary."[11]

Here we come to a major issue about the large abstract organizations that permeate our lives. What is their proper obligation to society as a whole and their own members?

Many believe that organizations do have a large responsibility for something other than their self-interest and, in the case of businesses, their profit. Inspired by the chaos and by the Gaia Hypothesis—the idea that the whole planet is a kind of self-organized life-form—some scientists, economists, and politicians have proposed that we ensure responsibility by assessing corporations a charge proportionate to the identifiable stresses each one puts on the environment and social fabric from which it profits.

The argument is that these stresses will have to be paid for by society sooner or later in the form of toxic-waste cleanup, unemployment, welfare, or municipal decay if the company abandons its community. The tax would be a fiscal acknowledgment of the feedback loops that link the corporation to the world around it but that corporations have been largely allowed to pretend don't exist. If corporations had to more fully acknowledge their connectedness to the reality in which they operate, wouldn't they make their products, conduct their business, and interact with their employees in a different way? And wouldn't the workers in these organizations feel that their work counted for something more than making money for strangers or serving the abstract demands of a bureaucracy?

The problem is that the "eco-audit" is a mechanical solution and would be meaningless without a change in the consciousness of society. If it were imposed externally, without that internal transformation, not only would loopholes be exploited but the intentions and spirit of the idea would be undermined. The point here is not the practicalities or politics of these proposals but the way in which they highlight the difference between the competitive, fragmentary, dominance view of reality and the chaos, open-systems view with its sense of inherent responsibility.

But let's stop at this point and allow chaos to play the trickster. Perhaps it has appeared that chaos is telling us our problem lies in the fact that we have created a reality where organizations and the individuals within them are fighting it out tooth and claw. But actually that isn't what chaos says. Chaos says the problem is *we think we live in that reality*. Because we think we live there, power, competition, and hierarchy come to dominate our psyches. However, chaos says that if we look closely at our current organizations, we'll see that something quite different from all that is going on inside—something that might even encourage us to change how we think.

Chaos reveals that real corporations are as much strange attractors as they are hierarchies. They are as much open, nonlinear systems—tied inextricably to the environment that gave them birth, subject to the fluctuations of that environment and the personnel flowing through them—as they are power centers. In fact, subtle influences and chaotic feedback are constantly at work within organizations.

From the chaos perspective, the real problem is that for a long time—perhaps since the beginning of "civilization"—human beings have imposed ideologies of hierarchy, power, and competition on top of their natural tendencies to collective creative activity.[12] We have magnified some elements of the collective process into the whole process. The result is that we now have a world full of organizations that are thwarting themselves and stifling the creativity and soul of the individuals who make them up. They are producing needless misery and psychological strife.

Complexity theorist Lynda Woodman and economist Brian Arthur have pointed out that newly formed organizations often have a flexible, searching, chaotic quality about them and a camaraderie among the individuals who start them. In the case of a business, this chaotic quality may allow it to burst successfully onto the commercial stage. However, after a time the organization falls prey to the grip of the standard "good business" assumptions and begins to petrify. Eventually, competition, hierarchy, and power begin to dominate the organization's activity. Negative feedback loops controlling the way things are done become reinforced, and soon the organization's strange attractor is reduced to a limit cycle. Arthur calls this "lock-in." Individual creativity is subordinated to the routines and ritualized beliefs of the organization. Many of these are so internalized that employees don't even realize they're there. People come and go within the company, but the "system" remains essentially the same. Individuals don't matter. As the organization becomes a powerful force in the marketplace, it also becomes less open to change. There's a reduction in the flow of information and degrees of freedom the company has to work with. The organization is like a bad heart with not enough internal chaos. Many companies fail at this point, succumbing to the emergence of new technologies or competitors who have more chaotic flexibility.

Business consultant Margaret Wheatley says, "It is strange perhaps to realize that most people have a desire to love their organizations. They love the purpose of their school, their community agency, their business. They have organized to create a different world, but then we take this vital passion and institutionalize it. The people who loved the purpose grow to disdain the institution that was created to fulfill it. Passion mutates into procedures, into rules and roles. Instead of being free to create, we impose constraints that squeeze the life out of us. The organization no longer lives. We see its bloated form and resent it for what it stops us from doing."[13]

Wouldn't it be great to participate in vital self-organized workplaces or live in self-organized democracies where our individual creativity generates the system and is, in turn, stimulated by it? As Whyte puts it, "What would it be like to grow organizations whose complexity arises from the cross-pollinating visions and imaginations of their constituent members."[14] To create these kinds of organizations within the context of 5 billion plus people may, in fact, be one of the great challenges facing humanity. It may well be that the planet's fate hangs on our ability to organize ourselves in ways that foster creativity instead of engendering alienation.

For individuals, creating the kind of organizations Whyte calls for would mean giving up some of the security we cling to in our traditional organizations (a security that is increasingly illusory, anyway). It would mean giving up our reliance on "leaders" as "heroes" who will save us or spare us the trouble of facing uncertainty. It would mean opening up ourselves and our organizations to the shocks, griefs, confusions, and mysteries that befall us by directly engaging the ethical, moral, and spiritual dilemmas of our activities. It would mean explicitly working with the tensions of diversity and divergences in points of view that are an inevitable part of collective activity but are now routinely turned into mere power struggles and the uneasy truces of compromise. In other words, it would mean being able to take the heat of creative chaos.

Here's a little story about one individual who did that.

The Dialogue Experiment

We'll call him Ed Brown. That isn't his real name, and we're going to change some of the details of his story because he asked us to. His name isn't important anyway, because he says, "Part of what I realized in all this was that we place too much emphasis on someone taking credit. The idea is that it's a process. That's what's important, not who gets stroked for what."

Ed's story begins when he joined a "dialogue group" some years back. Around the world different groups of people are coming together to explore the nature of group relationships. These aren't group therapy sessions. They're an attempt to understand how our individual and collective presuppositions control our interactions with each other and to explore the possibilities for collective creativity.

The physicist David Bohm, who devoted his last years to the investigation of dialogue, described it this way: Dialogue is "not an exchange and it's not a discussion. Discussion means batting it back and forth like a ping-pong game. That has some value, but in dialogue we try to go deeper . . . to create a situation where we suspend our opinions and judgments in order to be able to listen to each other."[15] This suspension is often less a willful act on the part of the group's individual members than it is an effect of dialoguing itself. Because there are so many diverse points of view flying around in a dialogue, everybody's opinions and judgments can end up getting suspended. Another dialoguer, painter and psychiatrist David Shainberg, called dialogue an "open process of making forms."[16]

One of the major ideas of dialogue is that people are tied to what Bohm called "nonnegotiable" convictions that underlie even their most casual disagreements. These nonnegotiables can't be reasoned out, but they may be suspended and transformed, as Ed discovered, through the process of collective creativity.

Ed's group consisted of about twenty individuals who met once a month in an apartment in a major city. They agreed to have no leader, no agreed project, no set topics for discussion. This stripped away the usual props that groups rely on, laying bare the issue of how individuals relate to the group.

"One thing I saw right away was that we wanted somebody to be the leader. We were uncomfortable without one. But nobody wanted it, or if anybody did we wouldn't let them keep it. Also we were all the time also looking for some structure and we couldn't agree on one. It was very frustrating. I think it's how a lot of people feel in groups. Always wanting to be heard, wanting the group to match your sense of things and feeling that it doesn't."

The group taped their sessions, and Ed learned that beneath the chaos of the frustration there was an order going on. "I saw that even though I felt I was never changing anybody's mind about anything, I actually was, and my mind was getting changed, too. It was very subtle. If you followed the conversation around, on one level it looked chaotic, but you could also see how people would pick up each other's words and ideas and internalized them somehow. It was pretty clear that we were all influencing each other.

"Sometimes you would argue with somebody and after a while you'd begin to see you really didn't understand what they meant. You were just reacting to the words. Once you got past the words, you realized they were saying something interesting. I also saw that I didn't really understand what I meant until people brought things out in what I said.

"Toward the end of the session, even though we'd talked about a hundred different things, most of the people in the group would seem to come to something. It was like we had created or discovered something in common but it was different for each of us. It was very peculiar."

One of the subjects that frequently came up in the discussion was whether what was happening in dialogue had any practical application. Ed decided he would test this question by getting involved in civic affairs in his suburban community and continuing the dialogue experiment in a "real, practical" setting.

Ed volunteered to work on a committee to renovate the library of his town and was selected as the committee's chairman. From the first meeting, he realized the group was headed for trouble. There were two factions on the committee, each one backing a very different kind of renovation solution. Some of the members of the committee were in-laws bitter toward each other from old wounds; there were also old towners and a couple of people who had recently moved in. The committee therefore contained a fair representation of the town's animosities and rivalries.

"Each of the factions wanted me to join their side. But one of the things I'd gotten from the dialogue group was that I didn't need to join the polarity. I was taking a chance that both sides would be mad at me, but what felt right was to listen to the factions and help each side make the best case they could for their position. Then I'd try to present one side to the other in as understandable a way as possible. It was more than just being neutral. There were a lot of tensions in our meetings—egos and power plays. The fact that somebody in the room wasn't taking sides and was actually interested in what people were saying confused the committee members at first but also freed them up. After a while they started to break out of their positions and ideas started to branch out until there weren't just two solutions but a whole bunch of variations. Soon even the most entrenched people were moving a little.

"But then we got stuck again and the old pattern re-formed. It was depressing. Things began to look really hopeless until one night one of these more neutral people suggested a solution that was a little different from anything we'd discussed before. To everybody's surprise—you could literally see we were surprised; our eyes jumped—we all liked it. In retrospect, maybe this idea was something we could have seen from the beginning, but we didn't. We couldn't see it before, but now we could. Probably we could because of all the movement we'd gone through. The whole context had changed. Suddenly, we were unanimous and we'd come up with a great solution for the town."

The solution was not a compromise. "Compromises are worse than defeat. They mean everybody feels cheated a little, or that you've given in to the power thing, winners and losers. This was much more exciting and interesting and satisfying to everybody than any compromise would have been."

Ed's authentic interest in the merits of the two sides and his skepticism that either side had the right solution provided a "subtle influence" that helped his colleagues suspend their polarities

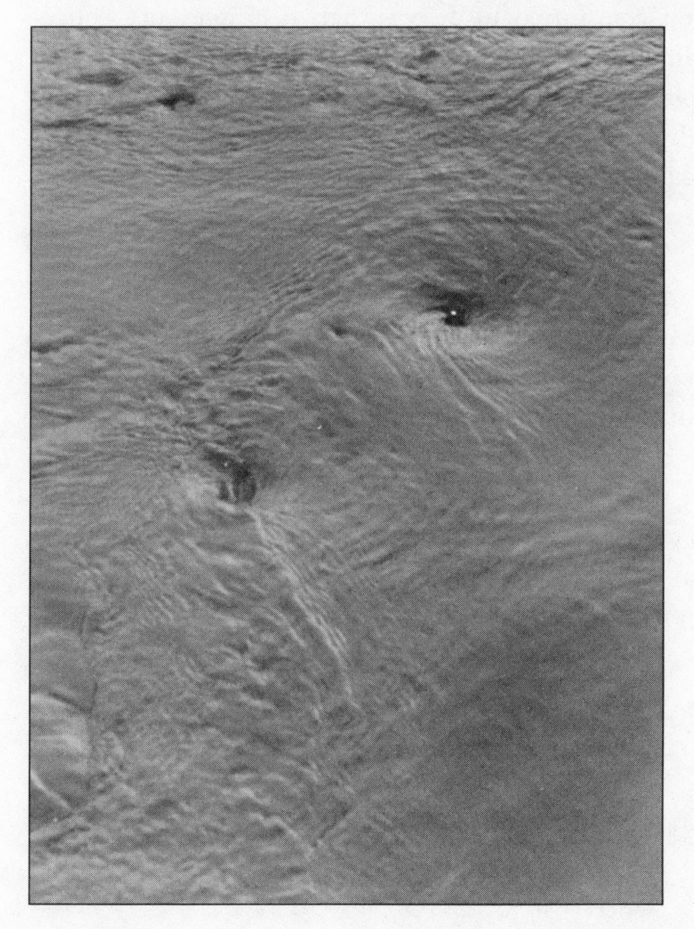

Photo by John Briggs

and nonnegotiable convictions long enough for something new to emerge.

Ed's story sounds something like the way many indigenous people work, such as the way Wilfred Pelletier's community fixed the council house roof.[17] Among the Iroquois, for example, the traditional council of chiefs was required to have the complete agreement of all its members on any decision. The Iroquois did not believe in majority rule. Their council sat for as long as it took to find a solution that everyone could agree on. Discussion was often vigorous and heated. Sometimes the councils lasted for days, weeks. In some cases, decisions were not made because no unanimous agreement could be reached. But when a decision was taken, it was one that everybody "owned" and felt committed to. It was their decision, both collectively and individually.

In our complex and problem-ridden mass society, we need to develop radically new understandings about collective action. What seems clear is that the problems of our collective world are such that no leader or system could ever resolve them. In fact, attempts to find solutions in that tried and untrue direction will undoubtedly lead to further complications.

The sad fact is that our organizations isolate and keep each of us apart as much as they hold us together. We have assumed that because individuals are essentially separate particles, collective action must be coordinated through these imposed external structures. But what if we dropped that assumption and allowed self-organization to create our communities? What if we intentionally forged our social solutions in the fires of creative chaos?

LESSON 4

Exploring What's Between

Lesson about the Simple and Complex

Is life simple or complex? Chaos theory says it can be both, and moreover, it can be both at the same time. Chaos reveals that what looks incredibly complicated may have a simple origin, while surface simplicity may conceal something stunningly complex.

The physicist Herbert Frochlich, a former classmate of Werner Heisenberg, one of the discoverers of quantum mechanics, knew about these processes. Asked about a theory he was working on, Frochlich complained, "It can't be correct. Nature does not work this way. The whole thing is too complicated." Some weeks later, after trying a different approach, he remarked. "It has to be wrong. The whole idea is too simple. Nature does not work this way."[1]

We have all experienced occasions when life's complexity confuses us. Trapped in a maze of alternative possibilities, a direct and simple decision becomes increasingly difficult to take. Yet chaos theory suggests that it is possible to discover a way out by engaging in chaos's dynamic dance between simplicity and complexity.

The Paradoxical Science

The very simple and the highly complex are reflections of each other. They're like the Roman god Janus, who is usually depicted as looking in two directions simultaneously and so possessing two faces inseparable from each other.

Insights about the paradox of simplicity and complexity occur repeatedly in art and ancient wisdom. In Dante's *Paradiso,* the poet journeys through a heaven full of infinite complexity and diversity, yet at the same time, all is contained within "one simple flame."

Paradox, a statement that appears simple yet acts to generate complex resonances within the mind, is a good way of thinking about how simplicity and complexity are entwined. The fourteenth-century philosopher Nicholas of Cusa depicted God by means of paradox—"the coincidence of opposites." Quantum theory, when describing the essence out of which matter and energy arises, refers to "the quantum mechanical ground state," which is both total emptiness—an absolute vacuum—and a plenum of infinite energy.

Paradox is central to the ways Eastern philosophy attempts to see the truth beyond our restrictive ideas of reality. In a famous passage from the Taoist Chuang Tzu, the master dreams he is a butterfly and then wonders if he is really a butterfly dreaming he's a man.

Fractals—the geometry of irregular shapes and chaotic systems—are a way of seeing and thinking about the complexity-simplicity paradox of nature. Trees and rivers, clouds and coastlines can be described by fractal geometry. The endlessly detailed Mandelbrot set, a portion of which is shown on the next page, is a typical example of a "mathematical" fractal. It comes as a surprise that this inexhaustible complexity was generated when a computer was fed with a very simple mathematical rule that was then applied or iterated again and again.

Mathematical fractals grow through a process of computer recycling, with the output from one cycle becoming the input for the next. At one level, the complexity of the fractal is a curious illusion, because although the figure's detail may be infinite, the way it grew was simple. This is also true of many of nature's forms and processes. For example, the complexity of a termite's nest is the result of the constant repetition of one simple action.

But incredibly complex, chaotic processes can also give rise to clear, regular structures, as when the chance fluctuations in a heated pan of water couple together into a regular pattern of hexagonal vortices.

In other instances, complexity and simplicity exist together

One layer of detail from the Mandelbrot set. To see some previous layers of detail for this particular image, turn to the next lesson. *Generated by Silvio Tavernise*

symbiotically in the same time and space. Electrons in a normal metal at room temperature behave like a chaotic gas of particles; they are free individuals colliding randomly. But give this "gas" an additional kick of energy and, like a plucked guitar string, it vibrates in the regular manner scientists call "plasma oscillations." Unlike the heated pan of water, the plasma isn't a random gas that suddenly self-organizes its state into a uniform oscillation. Both regular and chaotic motion are already simultaneously present. Zoom in on individual electrons and they appear to be banging into each other at random. Zoom out and see that on top of that random motion is a pattern of regular oscillations. The physicist's treatment of a plasma shows that the infinite complexity of chaos and simple order are indissolubly linked. Without the regular plasma order, it would not be possible for electrons to behave in a free, random way; similarly, the collective motion of the entire electron "gas" exists by virtue of the chaotic motion of individual electrons.

David Bohm, the scientist who created this plasma theory, saw it as an image of the way the complexity inherent in millions of free individuals, each one uniquely different, also produces a coherent society. Like the plasma vibration, society is a relatively simple and stable form that emerges out of the complex dreams, desires, and contributions of its members. Likewise, each individual, with a freedom of choice, is partially the creation of the society in which she lives.

A healthy society draws upon the energy and creativity of its members and at the same time provides them with values, ethics, and a shared sense of meaning. The Czech communist society described by Havel had swung too far toward simplicity. Individual creativity became stifled and the complexity inherent in human freedom minimized. At the other extreme lay Margaret Thatcher's Britain, with its emphasis upon a free-market economy. Here, individual freedom and creativity were elevated, but without sufficient attention being given to the way government and other organizations should manifest shared social responsibility. In the light of chaos theory, Thatcher's famous remark "Society does not exist" seems especially ironic. A healthy society requires attention both to the individual and the collective, the complex and the simple.

Intermittency: The Storm Within the Calm

Whenever interactions, iterations, and feedback are at work, simplicity and complexity constantly transform into one another. The situation becomes particularly striking when the simplicity and complexity alternate in what scientists call intermittency.

Nothing seems more regular than the length of the Earth's twenty-four-hour day. Right through the first decades of this century, astronomers established a standard of time by observing the transit of certain stars as the Earth rotated with respect to the night sky. However, with the introduction of highly accurate atomic clocks, it was discovered that the Earth occasionally jitters in its rotation. The passage of the Earth's "time" is not perfectly regular, but contains intermittent bursts of chaos.

The same thing happens with some electronic equipment. Certain amplifiers occasionally produce short bursts of static. This isn't caused by external interference, but results from nonlinear effects within the circuitry, producing periods of chaos. Sudden bursts of random behavior also occur in such systems as superconducting switches, prices on the stock market, nerve signals, and computer networks.

Intermittency not only means bursts of chaos within regular order, but also outbreaks of simple order in the midst of chaos. But it may take a bit of mathematics to see it.

Everyone knows that when rabbits got loose in Australia they rapidly spread across the entire continent. When a birthrate is such that two parents only produce two offspring, the birthrate is 1.0 and things are stable. In the case of rabbits, the birthrate was much larger than 1.0 and so the population grew exponentially.

But environments are finite; there is only a given amount of food to go around, and species have predators that catch and eat them. Just as one factor is causing a population to increase, another is pushing it in the opposite direction. In 1845, mathematician P. F. Veehulst, in an act that anticipated modern chaos theory, wrote down the simplest possible equation that contains these two competing factors. It is a nonlinear equation where our old friend feedback is at work.

Marshaling modern computers, chaos scientists investigate Veehulst's equation in populations with different birthrates. With a birthrate of 1.5 (two parents producing three offspring), the population grows to a steady value and maintains it over the years. With a birthrate of 2.5, the population overshoots the previous value, oscillates a little, and then settles down to the same value as with a birthrate of 1.5—the opposing factors of growth and diminution have again found their equilibrium. With an even higher birthrate, the oscillations continue for many more years before settling down. But at the critical value of 3.0, something curious occurs. The population no longer oscillates and settles on one equilibrium value; it now has two possible equilibrium states.

With an even higher birthrate, the population goes through a four-year cycle, high one year, low the next, then a new high, but not quite to the previous high value, then an intermediate low, and then, in the fourth year, back to the original high.

If we further increase the birthrate, the system becomes ever more complex.[2] With a birthrate of around 3.7, the entire system has no equilibrium and fluctuates unpredictably from one year to the next. This is the obverse of our pan of water. Rather than order being born out of chaos, chaos has been created out of regular order. Yet within this chaos can also be found residues of the simple. Look at the accompanying picture. It displays intermittency. The black bands are signs of order (relatively stable periods) appearing smack within the white wave of chaos. In these windows, the birthrate is stable again.

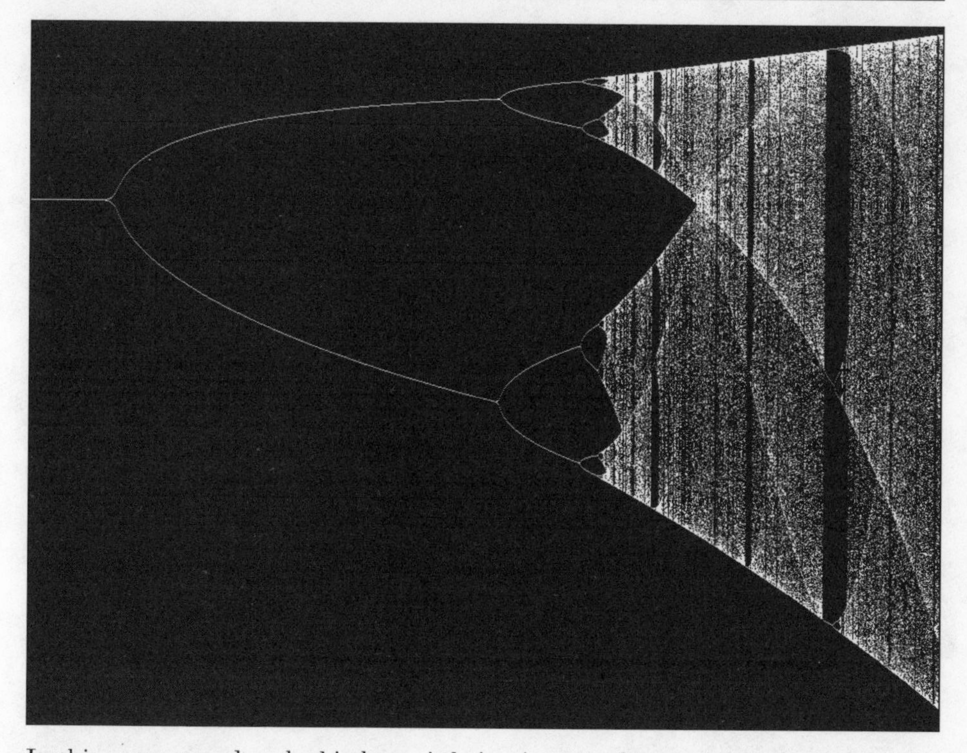

In this computer plot, the birthrate is being increased as you go to the right of the drawing. Where the single horizontal line on the left doubles into two lines, the population rate is 3.0. Where the waves of white begin, the rate is 3.7. *Generated by Silvio Tavernise*

Intermittency raises the interesting question: Does chaos emerge because regular behavior has temporarily broken down? Or is regular order really a breakdown of reality's underlying chaos? Do riots occur because the good order of society has failed? Or is a stable, calm society intermittency's manifestation of underlying chaotic complexity?

Many societies give intermittency an explicit role. It usually goes under the name of carnival or fiesta, an outburst of happy, creative chaos, a time for dressing up, eating, dancing, flirting, building bonfires, and general rule-breaking within otherwise ordered social norms. Such bursts of chaos allow the good order of society to continue throughout the rest of the year. Such societies understand a simple but complex truth that underlying the chaos of carnival is the renewal of important feedback loops that hold society together.

Intermittency is like a brief storm on a hot summer's day—bustle and noise that ends as fast as it began. Each flash of lightning means that nitrogen in the air is being converted into a water-soluble fertilizer that comes down with the rain. The storm brings a few moments of inconvenience, but it also renews the earth.

Sometimes chaos also bursts, uninvited, into our lives and can result in renewal or transformation. Intermittency is the unwelcome guest at a party. An irrational act, striking dream, or unfortunate coincidence challenges the normal order of our lives by asking us to give more attention to its nuances and subtle patterns. An unexpected illness or a child who gets into trouble can have the effect of cementing a family together. Too much stress makes people ill, but researchers have discovered that a little of life's chaos is necessary for the immune system to function efficiently.

Complexity and Chance as the Gateway to Order

Intermittency is a dark secret of the Universe discovered more than two thousand years ago. The Greek philosopher Pythagoras believed numbers were gifts from the gods, and mathematicians

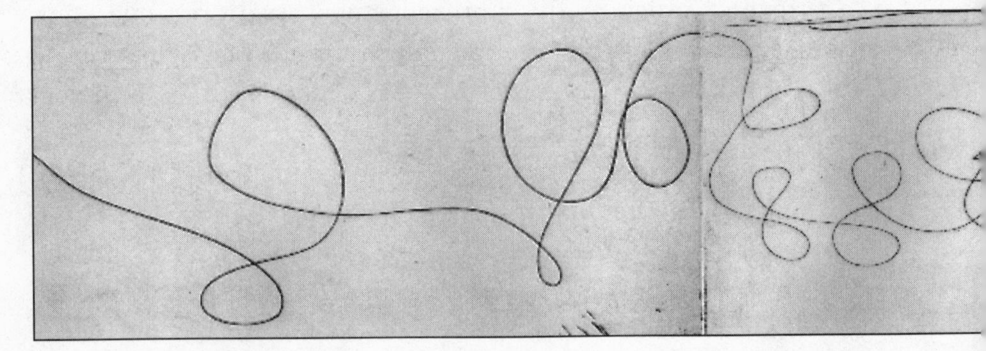

Detail from Landscape 1990 *by Nachume Miller*

today still repeat the aphorism, "In mathematics the numbers are God's, all the rest is man's."

Numbers are pictured as lying on a line marked off by the milestones of integers 1, 2, 3. . . . Between these milestones are fitted the rational numbers—numbers made out of ratios of the integers. For example, between 1 and 2 can be fitted numbers such as ½, ¾, ⅞, $^{31}/_{32}$, etc. Actually, an infinite number of rational numbers exist between any two integers. Moreover, between any pair of rational numbers, no matter how close they lie, exists an infinite number of other rational numbers. The Pythagoreans felt they knew everything about the numbers. There were no gaps, no holes in which to put anything else.

Then Pythagoras discovered his famous theorem about the square of the hypotenuse of a right triangle. He used it to calculate the longest side of a right triangle in which the two other sides are one foot long. To his horror, the result—the square root of 2—turned out to be a number never before seen in mathematics. It is an irrational number, one that cannot be expressed as a ratio of two other numbers. If we try to write down an irrational number, we never come to the end of it. Rational numbers, no matter how complicated they may be, are always finite or, like ⅓, repeat in a perfectly regular way (0.33333 . . .). But an irrational number is infinite; it has no internal order to indicate what the next digit will be. In a line that was previously filled with numbers, the irrational number creates its own gap and fits itself in.

The result was so scandalous that for a time it was suppressed by the Pythagorean brotherhood. In fact, it now turns out that the deeply significant numbers of the natural world, like the number pi, which relates the diameter to the circumference of a circle, are irrational. Irrationality is a form of intermittency within the regular number line. Irrational numbers are bursts of infinite complexity, of total randomness inside an otherwise regular system. Irrationality, therefore, lies at the very heart of both logic and the cosmos. Irrationality also exposes something quite curious about complexity.

We can begin with a simple system and allow it to develop in ever more complex ways so that its internal order becomes richer and richer, yet in the limit, when this complexity becomes infinite, it ends up looking exactly like chance and randomness—the opposite of any order. How can this be? Give a computer a simple rule and it will generate a rational number of a particular length. Make the rule more complicated and the number will get bigger and bigger. Yet it is always possible to detect an internal order. Give that number to a second computer and it will be able to figure out the rule by which the number has been generated.

But what happens when the rule becomes so complex that it requires many pages to write down? What happens if you need an infinite number of pages in order to write down the rule? Now the number is infinitely long and has no internal patterns or repetitions, so the second computer will work for years and years without ever discovering any hidden order or numerical pattern. A number lacking any internal order whatsoever is by definition random. For all practical purposes, the number created out of infinite complexity is therefore identical to a random number having no internal order. At this paradoxical limit, total chance and randomness become identical with infinite complexity. Push complexity too far and it becomes pure chance. Compress the simple and out bursts complexity.

Mathematics exposes one side of the paradox, psychology

another. No matter how chaotic and random life seems at the moment, we also sense that it contains some underlying order. People engaged in creative pursuits use chance—the odd spill of paint, an overheard fragment of conversation, the sight of a road sign—as germs and pathways to new forms. Chance events can offer a clue to some deeper pattern in our lives. The Swiss psychologist Carl Jung called apparently disconnected but highly meaningful coincidences "synchronicities" and suggested we should be willing to read these hidden patterns.

Synchronicities sometimes happen when we face an important decision or are in such a desperate situation that we're willing to stake everything on a last throw of the dice. Maybe synchronicities are always around us. It's just that in extreme situations we're more open to letting chance reveal patterns about hidden aspects of our lives. Pure randomness is the same as infinite information, but sometimes that infinite complexity of chaos divulges a clear and simple message.

Catching Ourselves in the Act of Simplification and Complexification

Chaos theory tells us that when life seems to be the most complicated, a simple order may be just around the corner. And when things appear simple, we should be on the lookout for the hidden nuance and subtlety. If the complicated turns out to be simple, and vice versa, does this mean there is no objective assessment of complexity? Are complexity and simplicity totally subjective ideas?

Chaos theory's answer is that complexity and simplicity aren't so much inherent in objects themselves, but in the way things interact with each other, and we, in turn, interact with them.

The British painter Patrick Heron writes, "The actual 'objective' appearance of things is something that does not exist—or rather, it exists as data that is literally infinite in its complexity and subtlety. What assuredly floods in upon the retina is an amorphous chaos of visual stimuli into which the human eye learns to inject a favored order of some sort or other."[3]

Perhaps rather than saying that the eye "injects" a favored order, we should say that the eye and brain, in their constant movements, interact with the activity and transformations of the world to abstract and "draw out" significant features. These features constitute the order we see. Our potential interactions with the world are so incredibly complex that our brains have evolved many strategies of abstraction and simplification. These strategies change over time. Nowadays, for example, we increasingly use a digital strategy. In many venues of our modern technological society, if a subject can't be digitized it isn't considered a subject. Some educators have warned that although computers can extend our grasp of the world, we should be careful not to rely on them so much that we ignore the many dimensions of reality that can't be computerized. Although the digital reality may seem complex, in fact it is a considerable simplification of the real world.

Similarly, science, in its desire to understand nature, has employed strategies that objectify and divide reality into manageable chunks for study. Using mathematics as a filter, science abstracts and simplifies nature, and this has made science itself possible. The approach led to Newton's celestial mechanics, molecular biology, the chemistry of synthetic materials, relativity, quantum theory, and now chaos theory. But the reliance on mathematics as an abstracting tool means science can only deal with what is quantifiable, numerical, and measurable. Thus scientific progress takes place at the expense of nature's qualities and unquantifiable values. This gives science a built-in tendency toward fragmentation and oversimplification.

To emphasize this point, the biologist, physician, and "biology watcher" Lewis Thomas suggested with irony that scientific researchers should direct their energies toward trying to understand just one organism completely—a protozoan living in the gut of an Australian termite. Thomas argued that if all the world's laboratories and supercomputers were focused on this one simple organism, we would soon realize that we can never know enough about it. What appears simple when abstracted as one isolated organism in a file folder of other similar organisms becomes endlessly complex the more we engage the organism's details. Because chaos tells us that everything is ultimately connected to everything else, gaining really deep knowledge about the protozoan would require understanding its connection to the entire history of evolution and the entire dynamics of its environment.

What is true of a protozoan is true of ourselves. To fully know oneself would require, in effect, understanding the whole Universe. Meanwhile, the literature of self-help and popular psy-

Nature makes forms by artfully meshing simplicity and complexity, as with this fern. *Photo by Lawrence Hudetz, 1989*

chology takes a simplistic approach. It is generally based on the premise that an independent "self" exists that can be identified, analyzed, reprogrammed, and improved. Yet if we really look for this self, what happens? The more we try to pin it down, the more we encounter our complex nonlinear interconnections to what is "outside" the self. The Buddha asks whether our ego exists in our sensations, in the forms of our bodies and brains, or somewhere in a chain of causes and effects, action and reaction. The more you seek this ego, this simple, essential self, the more it vanishes as an independent entity and becomes only a mirror that reflects the world.[4]

What is true of the self is true of "the other." With careless ease we simplify and stereotype individuals who are members of different groups. A stereotype—whether it's a positive or a negative one—is a cartoon exaggeration of traits or behavior that are assumed to be a characteristic of everyone in the group. In a stereotype, subtlety and individuality are lost. Mary isn't encouraged to study higher mathematics because "girls can't do math," or a patient is afraid to ask questions of her physician because "doctors think you don't trust them if you ask."

The environmentalist Barry Lopez, writing about the Arctic environment, emphasizes the way biologists are placed under pressure to produce statistics and computer models—scientific stereotypes—of the animals they study. Industries encroaching on the fragile environment are concerned about the effects on wildlife and demand that biologists provide a simplified description of a "standard animal." Lopez says, "Many Western biologists appreciate the mystery inherent in the animals they observe. They know that although experiments can be designed to reveal aspects of the animal, the animal itself will always remain larger than the sum of any set of experiments. They know the behavior of an individual animal may differ strikingly from the generally recognized behavior of its species, and that the same species may behave quite differently from place to place, from year to year."[5]

Our ability to simplify and make abstractions allows us to exercise a measure of predictability in our encounters with individuals and the environment. But when our simplifications lead us to idealize or denigrate others, we're in danger of losing touch with the reality of our actual connections. Perhaps one of the reasons that we experience such secret satisfaction in feelings of anger and hate is that they seem to make the world simple and clear-cut. Hatred projects the other as the enemy, offering us the illusion that if we could just eliminate the other, major problems would be solved. Feelings of love are more subtle and complex. In love, the other's depth and uniqueness are appreciated.

During a war, projection and stereotype operate with maximum force. The enemy is simplified into an evil brute, and the right and virtue of our own side are exaggerated. These stereotypes grip the minds of civilians and soldiers alike. They act to cement the nation and excuse the violence of war. But even in such dire circumstances, stereotypes can be broken. Soldiers taken to the edge of death in the service of the stereotypes sometimes transcend them.

In "The Man I Killed," Tim O'Brien's story about the Vietnam War, the narrator stares intently at a Vietcong soldier he has killed on a jungle trail. While his buddies cover the dead man up, the narrator's mind meditates upon the soldier's terrible wounds and imagines details about the man's life. Through this flowering of imagination and nuance, the narrator merges with his enemy, at the same time recognizing the impassable barrier of life and death that now separates them.

In a society so apparently awash in "information," data, and entanglements of all kinds, simplifications captivate us. We inhabit a television environment that packages the real complexity of life such as human interactions, social dilemmas, and the actuality of nature into sound bites and images that evoke simplified emotions. The image in the commercial of the refreshing stream connected to the name-brand beer, the somber music accompanying a montage of photos that recall the life of a dead celebrity—we often fail to realize that emotions can be just as stereotypical as ideas; in fact, they're two sides of the same mental process.

Simplifications are regularly used by demagogues to marshal our loyalty or fear so as to give the demagogue power. We constantly deceive ourselves about the grip that stereotypes have on us. Someone may actually believe he harbors no racial prejudice and then tell or laugh at a joke that centers on a racial stereotype. Such jokes create a false sense of community, an *us* as opposed to *them*. It's a way of blurring the real differences between the joke teller and his audience by exaggerating the differences between them and "the other" who is being caricatured. Nobel Prize–winning novelist Toni Morrison argues that the high levels of prejudice found against American blacks among recent immigrants to the United States has a similar function. Morrison believes the American social climate encourages the newcomers to buy in to the negative stereotypes of the black—America's historical "other." By agreeing with the stereotype, newcomers offer proof that they have accepted the majority culture.

Difference, which is a form of complexity, can engender feelings of apprehension and uncertainty. We may simplify those differences into something awe-inspiring, creating celebrities and heroes, or stigmatize them into negative stereotypes.

We should be at least as wary of our simplifications of people as we are of their complexity and difference. In a very real sense, he who simplifies is the one who is simplified.

At one time or another we've all seen this happen: Someone makes a racial or ethnic slur, using a stereotype, only to realize that his listener is a member of the group being slurred. "Oh, I don't mean you, of course." In fact, that may be true. Probably the listener is someone the speaker knows in enough nuance and detail to realize he doesn't fit the stereotype. Nevertheless, the stereotype remains an overpowering reality in the speaker's mind.

At some level we probably all know that stereotypes hardly ever fit the particulars of individuals. But our habit of using them as if they do contributes to an atmosphere in which they dominate our thinking and distort our interactions.

Getting Beyond Projections, Stereotypes, and Dualities

Seduced by our own simple abstractions, we can quickly come to see the world through categories that blind us to the subtleties and the richness of the small things that bring alive the individuality of each encounter and the freshness of each day.

But the obverse is also true. We can be so overwhelmed by detail and complexity that we become unable to abstract the underlying meaning of a situation. As we have seen, simplicity and complexity are not so much inherent in objects as a function of the way things interact. In both cases, we should be asking ourselves if the apparent complexity or simplicity is inherent in the particular issue we are facing or if it is largely something we are projecting onto the situation. In fact, the act of asking this question may stimulate our creativity in quite unexpected ways.

When Archimedes was asked if a gold crown had been adulterated with silver, he was faced with something incredibly complex. The crown was easy enough to weigh, but to determine whether it had the density of gold, he would have to calculate its volume. How could anyone work out the geometry of something with such complex detail? Then it struck Archimedes that the answer was simple. All he needed to do was immerse the crown in a tub and measure how much water it displaced—the volume of displaced water would be exactly equal to the volume of the crown. A sudden change of perception had reduced complexity to simplicity.

It's also important to distinguish between confusion and complexity. Complexity is telling us something essential about our interactions with the world. Confusion is quite different. It is a warning system that informs us we are failing to see the essential simple within the complex or we are overlooking the ripples of nuances within the simple.

One of our most persistent sources of confusion arises from our insistence on parceling the world into dualities. Expecting things to be either simple or complex is one example. Chaos theory points us beyond simplicity and complexity, objectivity and subjectivity, my view versus your view, order and randomness, stability versus hypersensitivity, naked power versus subtle influence, control versus uncertainty. It transcends these and other dualities that underlie our thinking and pump energy into our stereotypes and projections. Chaos theory shows us that it is an illusion to separate the self from the other, and that it can be equally illusory to imagine a false or inauthentic merging of the self *with* the other.

The habit of duality is old. From earliest recorded times, we have tried to divide the world in a bifurcated way in the hope of discovering a fundamental basis for knowledge and belief. For some philosophers, the Universe was a plenum. For others, it was a vacuum. For some, reality was an eternal flux of endless transformation; for others, indestructible, indivisible atoms. We are told we must choose between free will and determinism, mind and body, continuous creation and a single big bang, order and chaos.

What if each pole already contains the other? How many people in fanatical pursuit of the good have ended up doing harm? The whole story of humankind's fall in Genesis turns on a duality: Satan's offer of knowledge to discern the difference between good and evil—and we've been struggling with this ever since. The problem is that our fixation on dualities causes us to obscure what is really going on. For example, is the evil, wrongdoing, and unfairness in society the result of individual evildoers and conspiracies of evildoers, as duality tells us? Or do these ills at least sometimes arise out of the activities of ordinary men and women who accept the stereotypes, slogans, and other simplicities of society while at the same time complaining that "it's all too complex"?

We want to escape the tensions of ambiguity and uncertainty, but the more energy we put into one pole of a duality, the more it takes the charge of its opposite. So what are we to do?

How can we be free of the grip of such dualities? Chaos suggests that irony, metaphor, and humor help to move us beyond duality into a new clarity of vision. Art, music, theater, and sacred ritual all employ rich, ambiguous forms to escape from the trap of duality, as do the disciplines of many of the world's wisdom traditions. For example, the Sufis, or Moslem mystics, often employ a subtle form of humor to foster insight, as in this story: "A man once asked a camel whether he preferred going uphill or downhill. The camel said, 'What is important to me is not the uphill or downhill—it is the load!'"[6]

Another way to get beyond dualities is a "dialogue" process such as Ed went through with his committee. The diverse opinions of a group create chaos and nuances in the polarities and allow for the appearance of creativity and self-organization.

Chaos theory, with its simultaneous acceptance of simplicity and complexity, order and chaos, One and Many, self and other, comes closest to the world's traditional wisdom as suggested by the Sufi story. Chaos invites us to adopt new strategies of life, to walk the tightrope between oversimplifying choices by ignoring subtlety and overcomplicating direct action and clear decisions.

We are well adapted to live in this tension, for human beings have evolved to fit in both everywhere and nowhere. Other animals have discovered their particular evolutionary niche. Our human trick is to have no single trick, but to live within the gaps and explore many different kinds of environmental spaces. Rather than being the king of the jungle, we are the adepts of chaos.

Our brains have evolved to spot the patterns within complex and ever-changing situations, while at the same time uncovering the nuances within these patterns. Consider: A baby's first act is one of the most complex and creative things human beings ever do. The infant learns to recognize its mother's face within an ever-changing flux of appearances. It discovers the essential pattern from among many other patterns, at the same time figuring out those tiny changes of expression that indicate the mother's emotions.

Our survival as infants and adults depends upon the brain's ability to abstract patterns. Yet this great skill works against us when we get stuck and project simplicities instead of attending to differences. History is filled with examples of prejudice, stereotype, and plain stupidity, where people have grabbed on to a simple idea that works well in one context and shoehorned it into situations where it doesn't fit. Just to take one of the countless possible examples: Large-scale mechanized farming methods that work relatively efficiently for industrial nations can't be exported to the Third World without producing enormous social disruption.

The brain has a nasty habit of clinging to its simplified way of framing something so that after a time the frame becomes the reality. But we shouldn't despair over our faulty projections, stereotypes, and habitual prejudices. Chaos theory tells us we were also born with the power to overcome them. At every turn, we meet our natural ability to detect the movement of small sensations beyond dualities. For example, the language we speak is perfectly adapted to encompassing that vast range of orders from simplicity to unlimited complexity. We can formulate explicit instructions for making a meal or write poetry filled with ambiguity, metaphor, and paradox. By applying the "art" of the simplicity and complexity paradox, we can touch the force of life that flows into and beyond our abstractions.

LESSON 5

Seeing the Art of the World

Lesson about Fractals and Reason

If you have ever lain on your back to watch big puffy cumulus clouds boiling across the sky, if you have stood on a coastline enveloped in the sight of the ocean swelling and uncoiling breakers against the land, if you have ever contemplated the mountains, then you know.

There's something revitalizing and deeply fascinating about the recurring and ceaselessly variable patterns of nature. Perhaps we stop to marvel at the way a network of erosion has etched itself into a hillside or, on a vaster scale, sculpted the intricate chasm of the Grand Canyon. Or we pause to appreciate the sensuous angles of tree branches; the exhilarating puffs and bursts of wind on a blustery day; the wild, shifting shapes of fire; the spatters of mold and lichens on the face of a cliff; or a dark night's crystal scatter of stars. Nature's patterns, at once familiar and unexpected, inspire us, satisfy us, sometimes terrify us. Poets, mystics, and everyday travelers on Earth turn to these patterns for solace, for a sense of continuity, for a glimpse of the divine mystery.

Nature's patterns are the patterns of chaos.

"Fractal" is the name given by scientists to the patterns of chaos that we see in the heavens, feel on earth, and find in the very veins and nerves of our bodies. The word was coined by the mathematician Benoit Mandelbrot and now has wide use in chaos theory, where fractals refer to the traces, tracks, marks, and forms made by the action of chaotic dynamical systems. Natural fractal forms include the crack in a rock ledge left by an earthquake or frost heave, the dendritic web of a river system, the once-only shape of a single snowflake.

Mathematicians have imitated these natural fractals using various kinds of nonlinear (feedback) formulas. Although infinite in their detail, mathematical fractals lack the subtlety of their natural counterparts. Nevertheless, they have brought scientists closer to visualizing the real movement of chaos that makes natural fractals possible.

Natural Fractals and Ourselves on the Coast

The classic illustration of a natural fractal is a coastline. Mandelbrot introduced the idea of fractals in a paper that asked an elegantly simple but fiendishly complex question: How long is the coastline of Great Britain? His answer provided some wondrously curious glimpses into the landscape of chaos.

Imagine Britain from a satellite distance—Britain on a world map. Bend a thread around the craggy line of the coast and then hold it against the scale of miles on the map. How long is the coast? The answer seems simple. Now repeat this procedure on a national map that has more detail. On this new map, we see more of the bays and indentations of the actual coast. Measuring our thread against the scale on this map, we find the coastline measures longer. With a highly detailed maritime chart, it measures longer still. Now try it on foot with a piece of rope and a tape measure,

Chaos fashions the ever-transforming, gnarled, fractal shape of the coast. *Photos by John Briggs*

making an effort to encompass every twist and turn. What about going down to the twists and turns on the molecular or atomic level? By this logic, Mandelbrot arrived at the surprising conclusion that the coastline of Britain must be infinite. We might add that not only is the coastline infinite, but because it is continuously being eroded, it is an infinity that is constantly changing. Mandelbrot also discovered that every coastline, from the smallest desert island to the Americas themselves, has the length of infinity.

A coastline is produced by the chaotic action of waves and other geological forces. These act at every scale to generate shapes that repeat, on smaller scales, a pattern roughly similar to the one visible at the large scale. In other words, chaos generates forms and leaves behind tracks that possess what chaos scientists refer to as "self-similarity at many different scales."

The shape of a particular tree—which is produced by all the interlocked chaotic dynamics of the genetic program in the seed and the flux in the environment, including available sunlight, weather, disease, soil conditions, the position of other trees, and so on—is mirrored at several scales. The trunk forks into branches, the branches fork into smaller twigs. Twigs contain leaves, which themselves repeat the dendritic pattern in their veins. In its large-scale shape and in its small details, the tree is a self-similar record of the moment-by-moment, unpredictable flow-through of the chaotic activity that created it and sustains it.

That record contains not only what is similar about the different elements of the tree, it also contains what is absolutely unique about each element and combination of elements. Trees of the same species standing together in a grove each have a uniqueness that makes us stop and say, "Look at that tree over there. Isn't it beautiful?" In the angles, turns, and rhythms of its trunk and branches, in its patterns of lichens, moss, and disease, in countless other details, we are glimpsing a dynamic picture of the individual tree—and its life in the flux.

For clarity's sake, let's stipulate that the term "self-similar" includes this idea of individual differences and uniqueness as well as similarities. As we'll see, there's a vast range of fractal self-similarity that is possible both in the forms of nature and in human consciousness. In some fractal forms—particularly the ones generated on computer screens by mathematical formulas—the self-similarity is somewhat mechanical. In other fractals—fractals in nature and art—what is self-similar is infused with what is different in ways that defy description.

A Chan Buddhist text says, "One particle of dust is raised and the great earth lies therein; one flower blooms and a universe rises with it."[1] The poet William Blake echoes the Zen text with his instruction in "Auguries of Innocence": "to see the world in a grain of sand, and eternity in an hour." Fractal self-similarity is the chaos version of this ancient and poetic truth.

Start to notice it and self-reflecting patterns of self-similarity become a transforming vision, subtly changing our experience of order in the world. Look up at a starry sky. If we penetrate into it by magnifying a small portion of the apparent empty spots with a powerful telescope, we see that the gaps between the stars are filled with stars. The deeper we magnify, the more we see that the stars have gaps and the gaps have stars. Each magnification is both a repetition and a revelation of something we have never seen before. In our minds, a sense of recognition is balanced with a sense of discovery and freshness.

In our microscale existence, each of us, like the tree, is a unique representation of the world that made us. Perhaps it is fitting that in the first weeks after conception, a fetus passes through resemblances of a fish, an amphibian, and other mammals, traversing a microhistory of the chaos of evolution until it finds its own form and face. Biologists have even discovered that the mitochondria inside our cells contain remnants of a much earlier state of evolution in the form of DNA closer to that of bacteria than to the DNA found in the nuclei of animal cells. Native Americans along the Pacific Coast of North America fashion masks that contain hinges with one face beneath another in layers: killer whale, wolf,

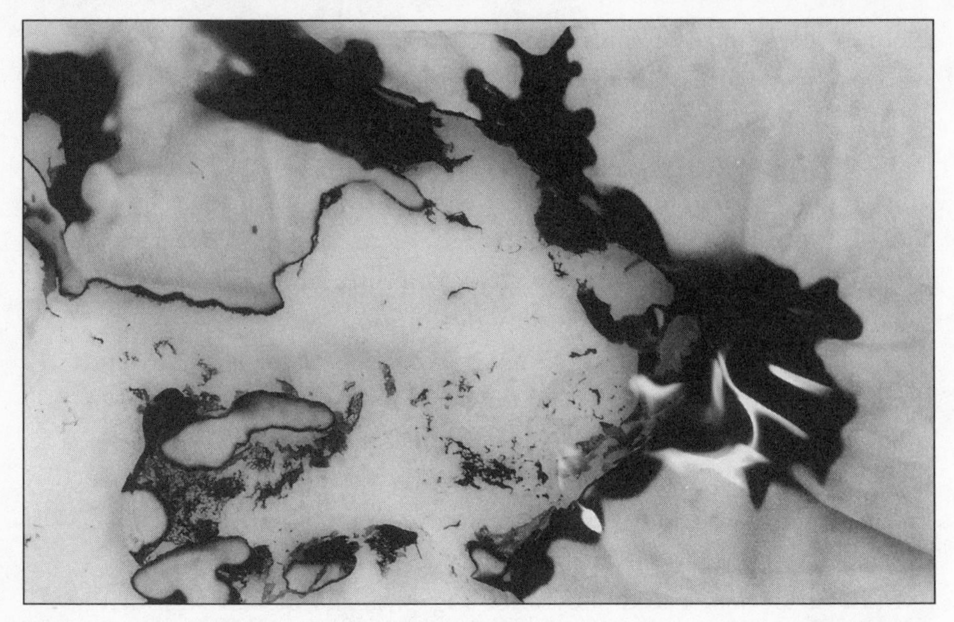

As the chaos of fire burns, it creates a fractal shape, reminiscent of a coastline. Each fire that burns and each moment of a fire is unique, yet there is a deep self-similarity crackling here as well. *Photo by John Briggs*

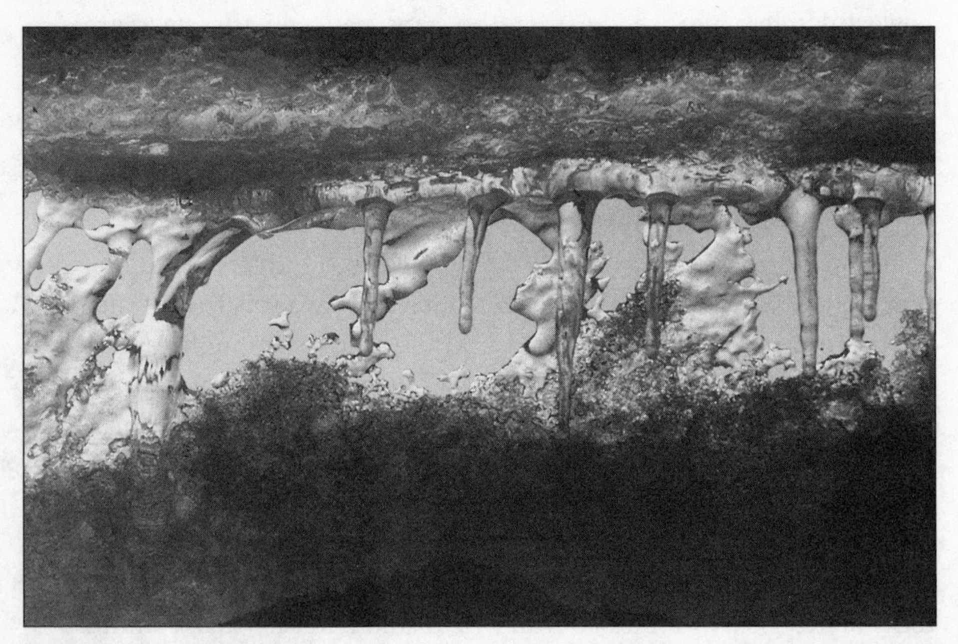

Coastlines appear to be all over the place in this photo of ice melting in front of a window. Notice the self-similarity of the ice shapes, but also see how each ice shape is wonderfully unique, a record of the chaos acting at that small place in the window. *Photo by John Briggs*

Erosion produces fractal shapes that record the chaotic processes at work on the rock. *Photo by Richard Halliburton*

A diseased tree blossoms a self-similar form. *Photo by John Briggs*

Cocoa drying at the bottom of a glass mug created a fractal pattern. *Photo by John Briggs*

On a moonless night, artist Susan Derges placed a six-foot piece of photographic paper at the bottom of a riverbed and exposed it using a flashlight. In her darkroom, what emerged was a snapshot of the fractal movement of currents. *Untitled photogram from River Taw series, 1996, by Susan Derges*

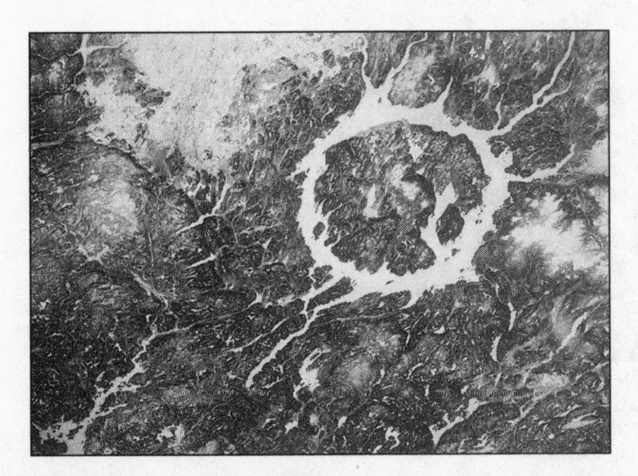

Erosion and weather has turned the ring of this asteroid impact crater in Canada into a fractal. *Photo by NASA*

Chaos theory has discovered that turbulent shapes such as clouds display patterns of self-similarity that the human brain can recognize. *Photo by John Briggs*

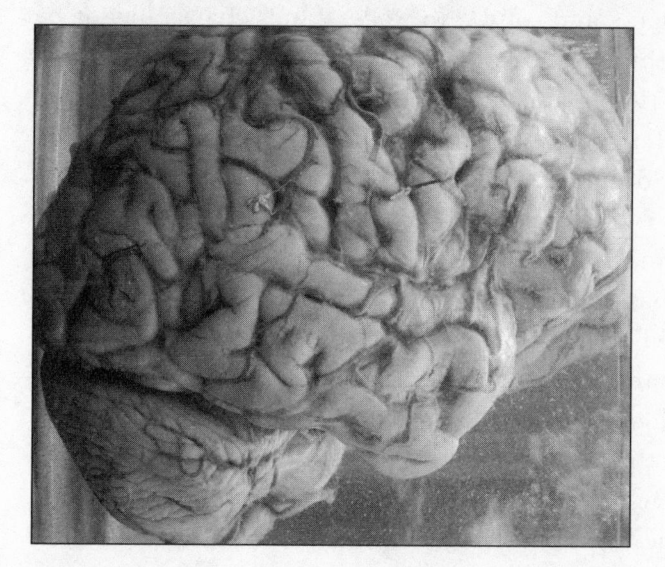

The human brain is also a fractal. Its craggy, self-similar folds are produced by genetic programs coupled with the chaotic, self-organized movement of neurons into place within the fetal skull. In fact, chaos dictates that even genetically identical twins will not have the same tangled neuronal wiring pattern within their brains. Each brain is its own unique fractal. *Photo by John Briggs; brain courtesy of Western Connecticut State University Biology Department*

eagle, man. The masks are used in ceremonies that honor and connect to the many beings reflected in our bodies and consciousness, the many scales and loops that echo the chaos of our self-organization.

The fractal self-similarity between microcosm and macrocosm (which includes the *dis*similarity of uniqueness and difference) is a product of all the complex internal feedback relationships going on in a dynamical system. Paying attention to the fractal features of reality is a way of glimpsing the mysterious, unpredictable movement that creates the world and holds it together. For a scientistic society, this is a new way to see.

In the culture that surrounds us, hype, hyperbole, and melodrama of all kinds have tended to desensitize us to the simple fact that an image of the essential mystery and order of life can exist as much in a small corner of the backyard as in the grandeur of a sweeping panorama.

It seems appropriate that Mandelbrot was able to show that the fractal phenomena of the natural world take place in between our familiar three dimensions of length, width, and height (represented by line, plane, and solid). To understand what is meant by the idea of something happening between dimensions, picture an ordinary piece of letter paper. Let the paper represent a plane of two dimensions, length and width.[2] Crumple the plane up into a ball. How many dimensions does it have now? It's not quite a sphere, but it's no longer a plane. In its folds and creases, it's an object existing somewhere between two and three dimensions.

Similarly, a coastline is unlike an ordinary one-dimensional line. It's crumpled and wrinkled to such an extent that it passes through vastly many more mathematical points on the surface of a plane than does a single straight line. This means, says Mandelbrot, that the dimension of a coastline must lie somewhere between the "one" of a straight line and "two" of a plane. (Britain's coastline has a calculated fractal dimension of 1.26, and all chaotic shapes—coastlines, rivers, trees, lungs—can be characterized by a fractal dimension.) One nineteenth-century predeces-

Once you begin to see it, fractal patterns are all around you. Here they show up in the distribution of flowers and grasses, as well as in the fractal bursts of wind that play over the area. *Photo by John Briggs*

Marie Hautem, a French photographer, is fascinated with living fractal patterns and tries to capture them on film. This picture of a horse recalls Gerard Manley Hopkins's famous poem "Pied Beauty": "Glory be to God for dappled things— . . . /All things counter, original, spare, strange;/Whatever is fickle, freckled (who knows how?)/With swift, slow; sweet, sour, adazzle, dim;/He fathers-forth whose beauty is past change:/Praise Him." *Photo, "Condense d'Espace Temps," by M. B. Hautem*

Fractals are everywhere. This one was on a wall in the poorest barrio of Manila, the Philippines. *Photo by Joe Martin Cantrell*

The fractal face of poet W. H. Auden, who wrote the fractal-friendly line, "Calling infinity a number does not make it one." *Photo by John Briggs*

sor of Mandelbrot imagined a line that was so twisted in its complications that it would pass through every possible point on a plane. Because the line also totally filled up the plane, he was forced to conclude that, in a certain sense, the line *was* a plane and that it paradoxically possessed both one dimension *and* two dimensions. The twisting and crumpling of natural objects and processes in our real world don't go that far. They exist somewhere in between dimensions. The fractal dimension of an object is a rough measure of how complex it is, the intricacy of its details. It does not tell us very much about the nature of the details and says nothing about their interactions.

The Aesthetics of Fractals

We have been taught to think of nature's "beauty" in terms of the sort of scenes found in travel films and on tourist postcards. The great Yellowstone and Yosemite areas were preserved as U.S. parks because they offered vistas in keeping with a concept of nature's awe-inspiring and sublime grandeur. The corollary of this has been that those parts of nature we don't think of as beautiful, pretty, or grand become disposable.

The aesthetic of chaos isn't about postcard beauty where the woods look like a city park or the grand views of Yosemite. It's about looking in detail at the real woods with dead trees fallen against each other, dense thickets, swampy sections, meadows containing poison ivy—a movement of things connected together in untold ways.

When nineteenth-century mathematicians discovered what are now called fractal shapes (the mathematical versions), they called them "monsters" and "pathological." This suggests our profound investment in idealized forms, an investment rampant in much of modern culture.

Natural fractal patterns evoke a recognition beyond the easy classifications of like or dislike, pleasant or unpleasant. We may not find an octopus a particularly attractive creature, but we may still grasp something essential in it. We understand that the octopus is in some sense us. For example, octopuslike forms are found in our bodies. At some level of our consciousness or society, we recognize tentacles. Perhaps the way we protect ourselves from our psychological predators is similar to the strategy of the octopus when it squirts out black clouds to create a "dummy" shape that allows it to escape.

A fractal aesthetic encourages us to explore the rich ambiguities of metaphorical connections between ourselves and the world rather than remaining only within the categorical abstractions that separate us from that world.

Our primal sympathy and appreciation of fractal forms contains our appreciation of the openness of forms fluctuating on the edge of life and death, living in the flow between structure and dissolution. Joseph Conrad called this our feeling of "solidarity . . . with all creation."[3] The unities we glimpse in fractal patterns aren't sentimental unities. They aren't unities that depend on a theory or even religious idea. They may even be unities that unsettle our theories and ideas. We may appreciate the fractal beauty of a war-shredded landscape or peer into the

Photo by Joe Cantrell

mirror of truth while reading a story about the grotesque conflicts of human nature.

Linda Jean Shepherd, an ecologist, says that in studying nature, science has traditionally ignored the "messy stuff, the monsters, the noises of nature." She calls these the feminine aspects of nature and believes that chaos theory helps us bring the feminine back into our exploration of the world.[4]

Math Fractals

Fractals came to the public's attention through the stunning abstractions generated on the computer screens by the famous Mandelbrot set. (For color examples, see the color section following page 104.) These images are plots of mathematical formulas. Mathematical formulas are, in turn, formalizations of the rules of logic. That certain formulas should contain a chaotic beauty is quite remarkable.

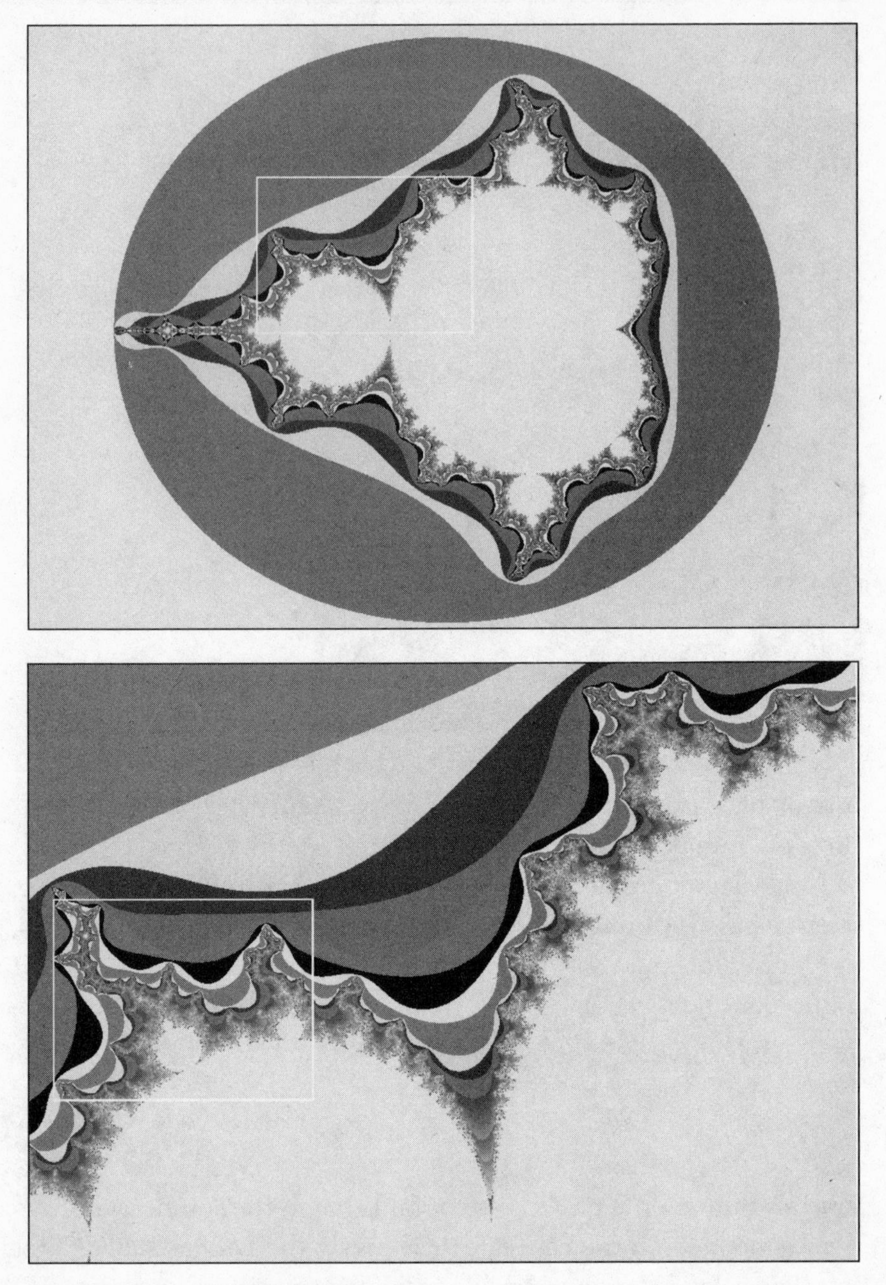

The first of these four images shows the entire region of the complex number plane containing the Mandelbrot set. The set itself is the tumorlike figure in light gray. When the numbers in this light gray area are "tested" by the nonlinear algorithm, they remain stable. Along the edge of the set, however, when the numbers are plugged into the formula, they spin off into infinity. Some power upward slowly, some quickly. Colors—or here, gray scales—are assigned to each of the different types of behavior the numbers exhibit. The numbers in this first image are

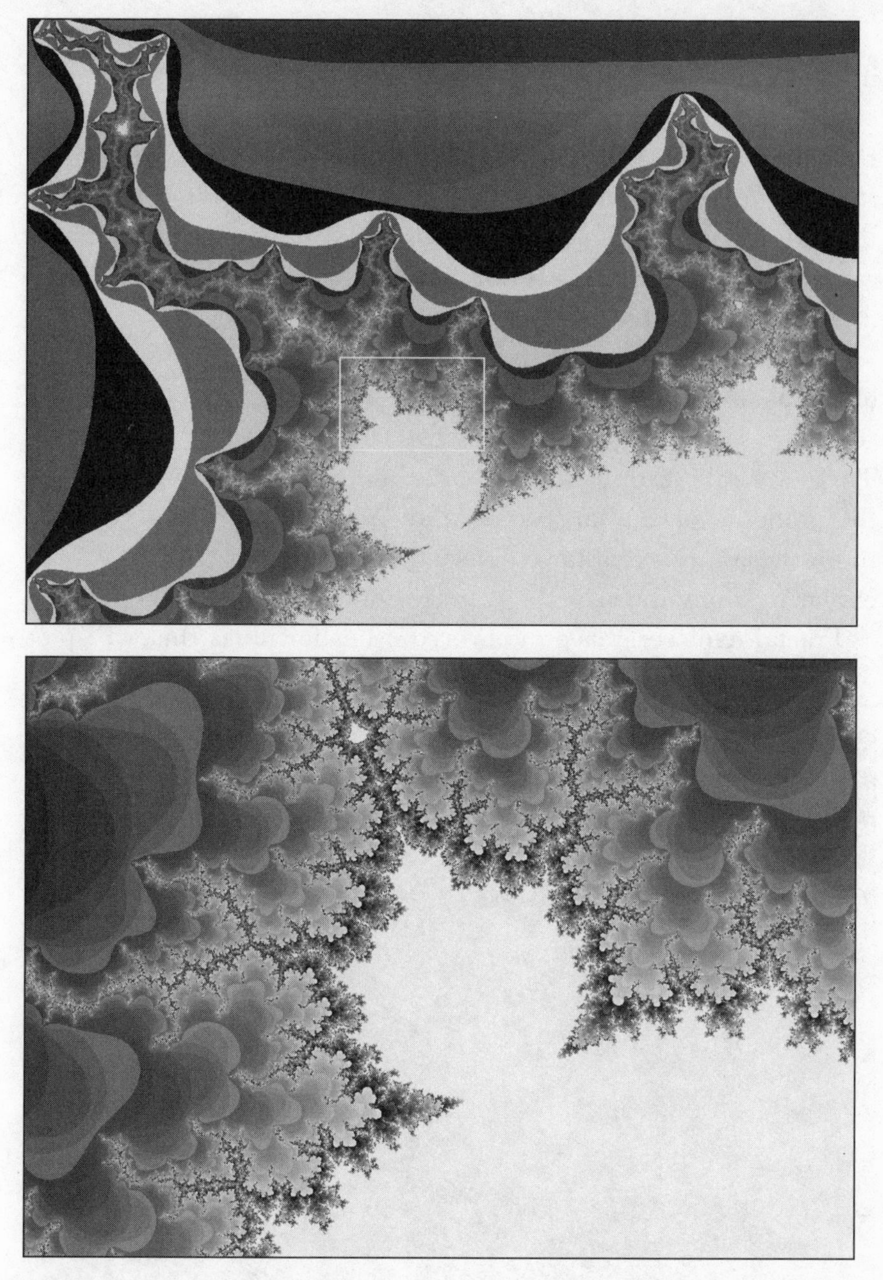

what might be called the large-scale numbers of the Mandelbrot region. When the formula is used to explore the numbers that lie between these numbers, the result is as if a camera zoomed into a close-up of some portion of the set. An area of zoom, indicated by the white box here, is shown in the second image. The third and fourth images are deeper zooms. Like the previous ones, these are accomplished by using the formula to plot an area of numbers between the numbers at the previous scale. *Generated by Silvio Tavernise*

The Mandelbrot set has been called "the most complex object in mathematics." It is a thicket of numbers located on one region of a mathematical construction called the complex number plane. By applying a simple nonlinear (feedback) formula or algorithm to the numbers in this region and then plotting their behavior as the formula iterates, mathematicians and computer "fractalnauts" can obtain stunning images that have a certain organic quality and a certain quality that resembles art.

As a zoom into the Mandelbrot set shows, this mathematical object possesses an incredible depth of fractal self-similarity. Even the large-scale image of the set itself is repeated at microscales. These are called mini-Mandelbrots. You can see one out ahead of the proboscis of the large-scale Mandelbrot. It's possible to zoom in on these minis and have a variation of the experience you had exploring along the original Mandelbrot edge.

Fractal explorers have discovered other formulas that will plot out fractal pictures on computer screens.

This fern looks realistic, but it's actually a plot of points laid down chaotically by an iterating nonlinear formula. *Generated by Silvio Tavernise*

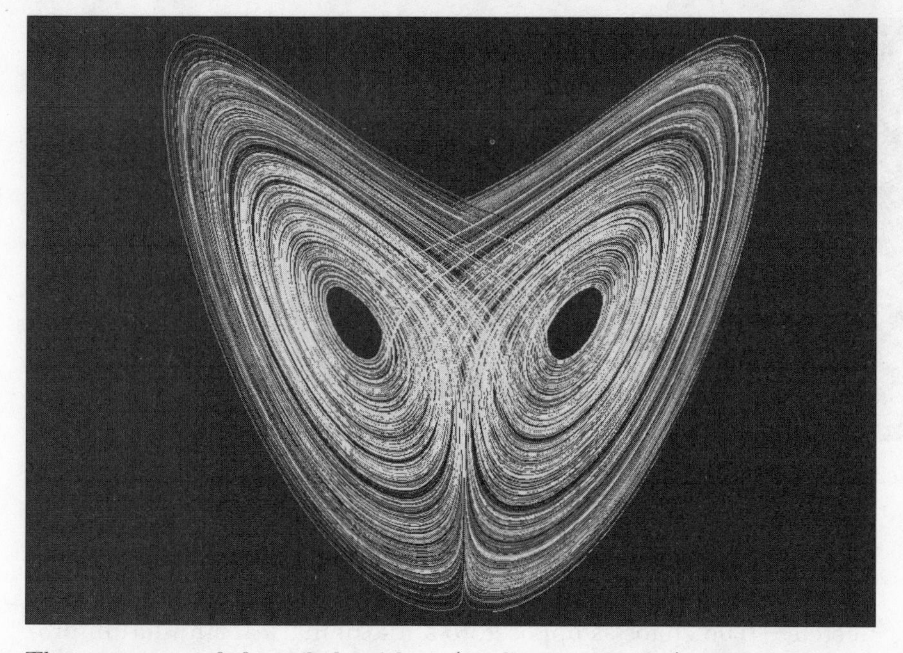

The same general class of algorithms (nonlinear formulas) employed to locate and plot the edge of the Mandelbrot set were used by meteorologist Edward Lorenz to model the weather. The isobars on weather maps have a fractal shape and look like the nested coastlines of islands. This image is another fractal shape connected to weather. It's called the Lorenz strange attractor. It's a graph made by coupling together the several nonlinear formulas of a weather model. The folded shape of the attractor represents the folding of feedback going on between wind speed, temperature, and pressure in a weather system. The successive lines moving around these folds in the plot indicate that the feedback going on at different scales is producing self-similarly at different levels of the weather system: For example, in the real weather the large-scale fronts moving across a continent are mirrored by the small-scale fronts we encounter moving down our driveway. *Generated by Silvio Tavernise*

Fractal principles have been used to construct imaginary mountain ranges and entire imaginary landscapes.

The companies that produce special effects for motion pictures regularly use fractals to create film reality.

Mathematical fractals are impressive, but after repeated viewing, the freshness of one of these objects fades. This doesn't hap-

Other State. *Copyright 1992 by F. Kenton Musgrave*

pen with the creations of nature, which emerge out of a holistic chaotic process whereby countless "parts" are subtly interconnected—true chaos as opposed to a mathematical simulation produced by repeating an algorithm. Consequently, natural fractals have an individuality, spontaneity, depth, and quality of mystery that no algorithm—even a nonlinear one—can reproduce.[5]

The Art Beyond Fractals: Joining Reason and Spirit

Throughout our history, art has been integral to the human experience of the world. From the time of Ice Age cave paintings through the Middle Ages, art was an expression of our faith that the Universe is spiritually coherent. Indigenous and peasant cultures lived, and many still live, surrounded by everyday objects—pots, knifes, animal skins—adorned with metaphoric self-similarities. They lived closer to the chaotic resonances of nature in which the spirit of life was revealed.

At its root, art contains nests of self-similarity. But the self-similarities of art, like those of nature, are deeper and far richer than those in the Mandelbrot set. The kind of "fractal" order in art goes far beyond anything mechanical—anything that can be reduced to didactic description. In fact, it's what defies our description that defines an artwork's greatness.

Becoming sensitive to an artwork's self-similarity is a little like becoming attentive to the way birds, squirrels, and chipmunks interact at your backyard feeder. After watching them for awhile you begin to sense that although there are repeating patterns, within these patterns something unexpected and profound is going on that keeps you absorbed.

Nature makes its fractal forms out of matter and energy. The material of art includes human consciousness, as well. Poems, paintings, and concertos are fashioned out of our categories of perception and language. Artists create discord within these categories by using irony, poetic metaphors (where unlike things are said to be the same), simultaneous harmony, and dissonance among musical notes and other analogous techniques. The concords and discords form patterns that are always surprisingly and significantly self-similar and self-different from each other, reflecting the curious mystery of our being in the world.[6]

For example, in a fugue, a simple theme is played in different voices and keys. The theme may be played upside down, backward, at different tempos, transposed up and down the scale. But the essential thing is not these somewhat mechanical rules for generating self-similar permutations of the theme. It's the way the rules are worked and broken, creating a flow that may remind us of the unexpected, transforming patterns flowing within emotions, thoughts, and nature. Listening to a great fugue is like listening to the inner movement of existence. Artists are tricksters, opening up possibilities and reflections. In the hands of a great composer such as Bach, the fugue becomes an organism transmogrifying and shifting within its repetitions and reflections. A work of art's simultaneous concords and discords subtly peel back our abstractions and reflex for making algorithms, showing us something that shines beyond them or lies glimmering within them.[7]

The rise of science and technology introduced a mechanical order into human consciousness that, along with other factors,

The Gothic cathedral was a great artistic structure of self-similar (and different) forms laid out in a tension and balance designed to evoke the appreciation of cosmic harmony. The divine light of God, *Logos*, the Being who was pure "reason" or rationality in the ancient sense, was made manifest in the space of the cathedral through the use of stained glass. *Photo by Richard Halliburton*

has tended to marginalize art. Though it may sound implausible, art was once viewed as a quintessential rational pursuit. Now we associate rationality with science and think of reason as the capacity to be logical, analytical, coldly objective, and detached. In earlier times, however, reason had another meaning. Apollo, the god of reason, was the patron spirit of the arts, especially music and poetry, and the god of beauty. Up through the time of the Middle Ages, rationality meant a mind capable of seeing the spiritual connections in things, the rhythms and delicate balance or "ratio" among subjects and objects. But art escapes from the confines of Apollo's harmonious reason as well. At the shrine of Pythia, the snake oracle of Delphi and Thrace, Apollo is portrayed along with Dionysus, the passionate and instinctive god of intoxication. It is as if these aspects emerge from the same source and are together inherent in the act of creation. Within creativity, order and chaos, design and chance,

Contemporary sculptor John Crawford says of this work, which is made of large blocks of wood, "The megaliths are for me a poetic tool for understanding the world and man's self-appointed centrality in that order, and the fractal structures are a less egocentric form for that same understanding. We are each the center of the universe and at the same time peripheral participants in immense patterns." Among other things, Crawford's forms generate self-similar/dissimilar tensions between the artificial and natural, the ancient and modern, wood and stone. He means for us to discover ourselves and the world in this gathering of odd blocks. *Photo by John Crawford*

planning and inspiration, endings and beginnings all go hand in hand.

Seeing the nuances and resonances in nature's fractals brings us back to the ancient gods with a chaos twist. Fractals and chaos allow us to add the rule-breaking Dionysus into our idea of what it means to be reasonable. If to logic we add harmony and to harmony we add dissonance, then to be rational is to be creative. In a world where we must make rational decisions that affect entire chaotic ecosystems, is it too much to think that we desperately

The Anasazi of the American Southwest built their villages into the walls of buttes, sensuously blending the Euclidean geometry of human thought-forms with the natural fractal shapes of the high desert. *Photo by Richard Halliburton*

need a new kind of rationality that includes not only our powers of analysis and logical deduction, but also our empathy and aesthetic response to the natural world?[8]

To be rational is to include the little "sensations" that Cézanne felt when he looked at a scene—that we all feel when we look at scenes. "Cézanne's doubt" should also be part of our analysis, an alertness in the chaos to the truth of the moment.

It is clear that our old form of reasoning, which takes the world as an exterior object to be analyzed, dissected, and controlled, simply isn't working in the context of the many problems that face our modern world.

To take one example, computer models suggest that it might be possible to heal the hole that our use of fluorocarbons has created in the atmosphere's ozone layer by using a fleet of large planes to spray 50,000 tons of propane or ethane into the South polar sky. Some scientists have theorized the hydrocarbon spray would set off a chemical reaction that could prevent the seasonal destruction of the ozone that protects us from the sun's harmful ultraviolet rays. This would be a clever technical solution, but would it be a genuinely "rational" one? If we let our new sense of rationality guide us, we see immediately that mechanically piling one technology upon the problems created by other technologies will only perpetuate the mind-set that is destroying our natural world.[9]

If we viewed our environment aesthetically, with this new sense of reason, as well as logically, analytically, and mechanically, wouldn't we want to live in it differently? And wouldn't it, in turn, be able to nourish us more deeply?

The American architect Christopher Alexander has studied towns and villages all over the world that possess what he calls "the quality without name," places where fractals and self-organized chaos flourish: "Places outdoors where people eat and dance; old people sitting in the street, watching the world go by; places where teenage boys and girls hang out, within the neighborhood, free enough of their parents that they feel themselves alive, and stay there; car places where cars are kept, shielded, if there are many of them, so that they don't oppress us by their presence; work going on among the families, children playing where work is going on, and learning from it."[10]

Alexander has found that where such communities exist, they were not created by a master plan, but through ordinary people unfolding their architecture out of the natural patterns of their lives. When a town or a building has this unnamable quality, it becomes "a part of nature. Like ocean waves, or blades of grass, its parts are governed by the endless play of repetition and variety, created in the presence of the fact that all things pass. This is the quality itself."[11]

He compares these natural patterns to the imposed mechanical patterns that dominate many of our high-tech lives: the rigid time schedules and deadlines of the organizations we work in, the planned developments made out of prefabricated materials, the clotted highways that bind us between home and work, the sharp divisions between work and family and leisure.

By using an aesthetic rationality, giving attention to the fractal world and creatively blending with it, wouldn't we feel—as complexity theorist Stuart Kauffmann puts it—more "at home in the universe"? The sensual self-similarity and difference of nature and art provide an inspiration, as Alexander says, "to be more alive."

So, in the end, we discover that chaos theory is as much about aesthetics as it is about science.[12] Chaos theory isn't art, but it points us in a similar direction: the direction we find in the healing images of nature, the direction in which lies our effort to contact the secret ingredient of the Universe we call spirit.[13]

LESSON 6

Living within Time

Lesson about the Fractal Curls of Duration

Time in our modern world has become our captor. The essence of time has been reduced to mere quantity, a numerical measure of seconds, minutes, hours, and years. We never seem to have sufficient time, yet when a little time is given to us, we waste it. Time's qualities have vanished. For us, time has lost its inner nature.

In other societies, time is an energy of the Universe, a river to be navigated, a bosom on which to find rest. In our postindustrial world, time has become mechanical, impersonal, external, and disconnected from our inner experience.

However, chaos theory shows that it is possible to reconnect ourselves with the living pulse of time. The last lesson of chaos was about living within the new dimension of fractal space. This lesson is about living within the new dimension of fractal time.

We'll begin with a simple story, one which crops up in different versions in many cultures. One day a monk, returning from the forest where he had been collecting wood, stops to listen to a bird. Its song is particularly beautiful, and the monk is held, entranced, for a few moments before continuing on his way. Arriving back at the monastery, he discovers new faces. While he was listening to the bird, all his friends died and a century passed by. Through entering fully into a single moment of time, the monk touches eternity.

The monk's story recalls Blake's assertion that it is possible to experience "the world in a grain of sand and eternity in an hour." Indeed, it resonates with the way creative people experience a time quite different from that measured by a clock.[1]

Time's Fractal Nature

As long as we believe time is a straight line, an arrow speeding from past to future, it's difficult to account for many of our inner temporal experiences. We usually dismiss them as delusions, dissociations, quirks of memory and perception, certainly not anything to do with the essential, physical nature of time.

Chaos theory replaces the line with an endlessly complex figure of fractal dimension. At every scale of magnification, the fractal reveals new patterns and intricacies. Chaos theory argues that there are no simple lines in nature. What may look from a distance to be linear reveals on closer examination the twists and turns and arabesques of infinite fractal detail. Other lines turn out to be not even continuous but composed of clusters of tiny discontinuities, and clusters within these clusters.

So what about time, that line we assume to run from past to future? Why should it be the only one-dimensional line remaining in nature? What if the linear time of our technological world is no more than a convenient delusion of our mechanistic world, concealing a living vibrant time within the interior curling details of a fractal?

This notion, that time has a fractal dimension, accords with our immediate experience. It gives us the door to enter into time's eddies and currents and to explore its turbulent vortices. In fact, we've probably already been there.

In the midst of a life-threatening accident, time can appear to stand still. Events happen in slow motion. We have a strange world of time to decide whether to brake or accelerate out of a potential crash. It is as if each event within the crash panorama is unfolding in its own individual time with its own special rate of being and movement.

This experience of time may not be so much an illusion brought about by a mind overcharged with adrenaline as a momentarily clear vision of just how things really are in the dimensions of time. In moments of crisis we temporarily disconnect from the mechanical time of the clock and enter into fractal time, experiencing its temporal nuance.

Listen to someone humming the first few notes of a familiar tune and the entire music shape seems to be born in our heads. It is something all of a piece. In one moment we have accessed a fullness of time inherent in those first few notes. Now try the experiment of asking someone to hum the same few notes, but this time with a second's interval between them. Now the notes remain just what they are, single sounds, each isolated in an island of time. The time of the music is no longer present to us; we hear no tune. The notes refuse to combine into any recognizable sonic shape.

When we are willing to enter into a fractal dimension, our experience expands into time. We explore time's nuances and act according to our own internal rhythms.

Cindy Warren loves to hike in nature. After a while, she begins to detach herself from schedules, timetables, deadlines, and appointments. She discovers that the bureaucratic, prearranged time line of her fast-paced culture has little to do with her own inner rhythms of life. The world she lives in divides the line of time into lock-step segments of duration with no room for fractal details. But, as she says, "When I'm watching a running stream, listening to the wind in the trees, or just looking at a frog trying to catch insects, I feel I've gotten into rhythms of time that have absolutely nothing to do with the numbers marching along my watch."

As we explore time's fractal details, microevents flood in on us full of nuances hardly noticed before, while at the same time we begin to sense the flow of vaster and slower waves of time—the movement of the Sun across the sky, the warming of the Earth, the growth of seeds, the aging of the trees. Those fractal dimensions of time are also curling and breaking inside our bodies. When the society we have created cuts us off from the deeper meaning of time, it robs us of our connection to the rhythms of life itself.

Brenda is a social worker whose clients are sometimes in pretty desperate situations. Yet some days she tells her boss, "The time's not right for me, I'm not doing any interviews this morning." Because she works for a Native American organization, her supervisors understand when she tells them the time she has with a person has to be just right, otherwise things can end up worse and not better. And so Brenda relies upon her inner sense of the quality of time, and when she and time are out of synchronization, she believes it is far better for her to go home and try not to make a mess of the world.

Breaking the Scientific Line of Time

There was a time when most people experienced life as Cindy and Brenda do. (In fact, the vast majority of the world's peoples still live in this way.) European peasants of the Middle Ages had no need for clocks. They read the changing patterns of stars in the night sky and knew the time to plant and harvest. They heard the cock crow in the morning and watched the first rays of the sun turning the sky pink. They worked until the hot Sun became too high in the sky for them to stay in the fields. They felt the cooler period of the afternoon and returned to work until the setting Sun brought them home again. They heard the cuckoo in spring and the distant sounds of the monastery bell announcing the offices of the day.

Gradually, this medieval consciousness of time began to shift. The church had once taught that time belongs to God, and so usury—lending money against time—was therefore a sin. But early in the fourteenth century, the first mechanical clocks began to appear on public buildings and time was well on the way to being secularized. The rise of banking, with its practice of loans and promissory notes, demanded a time in which the future could be anticipated and economically controlled. Time became abstracted from immediate human experience and reduced to a

number, something to be manipulated by an equation. How much profit will I make if my interest is compounded over twenty years? How long will it take me to pay off the principal of a debt? If I make 100 percent profit on a ship sailing to the East, is it worth locking up my capital for a whole year?

The only way this was going to work was if that symbol "t" for time was well behaved, or what mathematicians call "single valued." You can't balance the books if time is fractal or multidimensional. Time for an accountant can't keep folding back on itself, it can't be rich in its texture, it can't be layered.

The extent to which time was transformed into a commodity can be seen in the colloquial phrases of the English language: Time is something we spend or save, put aside or waste, and generally don't have enough of. This new vision of time ultimately made capitalism and the rise of international corporations possible. Time had become money and money was numbers.

Abstract, numerical time lends itself very well to physics. Here, equations only work when time is a number on a line. In physics, there's no room for bits or lumps of time; no grain of sand can be allowed to foul the watchmaker's universe. The time of science and commerce began to dominate consciousness. So bit by bit time became mechanical and monolithic.

Time today is much like a journey between two railway stations. We've left the station of our birth and are on the way to our final destination. We think of our life and living as whatever length remains of the track of time before we arrive at the last station. Instead of time being our companion and friend, it is what is being eaten up fast, just as the train eats up the track ahead of it. We desperately try to fill in the time remaining. We divide our journey on the track of time into months, days, weeks, hours,

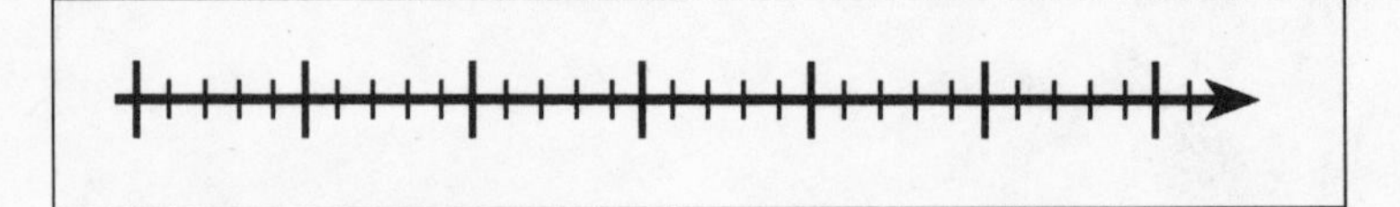

The track of mechanical time.

minutes, seconds, and, if we're working on a computer, microseconds. We always have to get a certain amount done within a given period of time.

This attitude is also reflected in our conventional view of history. History is a procession along a road whose milestones are battles, the deaths of kings, and the elections of presidents. Virginia Woolf suggested another sort of history, one in which women are engaged in continuous small acts of nurturing and holding society together. Woolf challenges our preoccupation with a historical time demarcated by dramatic "events" strung out along a line of time. She suggests that the real significance of time lies within the realm of subtle, human interactions and enfolded, multi-layered moments of human contact.

Fractals are self-similar, and so it is with fractal time. In a satisfying work of art, each portion of a painting metaphorically reflects the movement going on in the whole of the painting. In a great piece of music, such as a Beethoven string quartet, a fractal self-similar time is unleashed. Time expresses itself in the subtly changing tempos that are like water moving in a rock-littered mountain stream: time curls, spills, separates, flows around obstacles, merges, pools quietly, slips forward, flashes with light and darkness. Music invites us to be with each moment as it flows into new directions

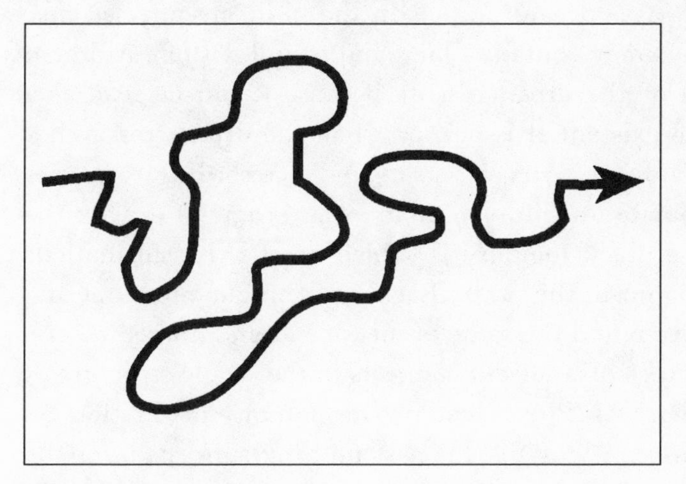

One view of fractal time.

In the moment the photograph was taken, each blade of grass was a fractal record of the time of the ice storm and the melting that followed. *Photo by John Briggs*

and detail. Yet always the paradox: Each moment of musical time is freshly yet subtly echoing the moments past and to come.

Psychiatrists say a dream unfolds in the brain in only seconds, yet those seconds may contain a long and complex story. A dream is a microcosm of the dreamer's life because it can be read as a reenactment of existential issues, or what one dream researcher called the individual's "survival strategies."[2] According to neurophysiologists, our brains never remember an event in exactly the same way twice. Each memory is subject to the transformations constantly going on in the brain. Each event in our remembering is both a new event and the same event we've remembered before. Each remembrance of an event connects to the whole structure of our consciousness. As Proust said in his famous meditation on time: "When from a long-distant past nothing subsists, after the people are dead, after the things are broken and scattered, still, alone, more fragile, but with more vitality, more unsubstantial, more persistent, more faithful, the smell and taste of things remain poised a long time, like souls, ready to remind us, waiting and hoping for their moment, amid the ruins of all the rest; and bear unfaltering, in the tiny and almost impalpable drop of their essence, the vast structure of recollection."[3]

Might it be that each "event," or even each moment, in our life is a fractal microcosm of our entire life?

Nature's Multiple Elastic Clocks

Everything from atom to cell, from tree to cosmos, carries its own internal clock that measures its individual passage of time, which is to say, the amount of process it has experienced. Chaos theory tells us that systems tend to self-organize, preserving their internal equilibrium while retaining a measure of openness to the external world. Something similar happens with time. Each element of a system possesses its own clock, its unique measure of the amount of internal process that is taking place with respect to the outer environment. In the self-organization of a larger system, the internal "clocks" of the smaller systems couple together.

Time's rhythms range from the fast ticking of the atom to the expansion of the entire cosmos. Time unfolds within the geological processes of Earth, the changes of the seasons, the life of a fly. Each system contains its own measure of time and, as systems connect into environments, time becomes ever richer and multidimensioned.

Every one of us is a multiplicity of internal clocks. Our cells have their own individual timekeepers that switch on and off various biochemical processes. Cells organize into individual organs whose internal clocks instruct them to secrete hormones and chemicals. These chemical messengers cause the time rhythms of various organs to couple together in the larger, self-organized system of the body. Some of the subsystem clocks operate on limit-cycle repetitions, the female menstrual cycle, for example, and high and low metabolism during the day, sleep and wake cycles. Other of our internal clocks—such as the many rhythms of the consciousness—are more open to environmental influence.

When the brain perceives something as a threat, a variety of signals override the normal cycles of the various organs. Adrenaline is secreted, which interrupts the regular heartbeat, speeding it up. Other secretions cause blood vessels to contract and move the major blood supply to the inner organs, reducing the effects, say, of a surface wound. Under immediate threat, the whole nature of our internal time changes so that the leap of an aggressive animal or oncoming car is reduced to slow motion.

The electrochemical activities of the brain are a measure of living time, a time that maintains a healthy balance between restrictive order and excessive chaos. Researchers have classified our various states of awareness—active thinking, dreaming, deep sleep, anesthesia, and even coma—in terms of the fractal dimension of the brain's electrical activity. All this suggests that the actuality of time, the time of perceiving and thinking, is quite complex and multidimensional.

As we have already seen, self-organizing systems sacrifice some of the individuality inherent in their components in order to give birth to the collective. Yet these hidden degrees of freedom are always present to animate the system. The brain operates with a multiplicity of internal clocks. We are conscious of some of these when we think out a problem in chess or try to explain why the economy isn't going to pick up. But others, such as the control of respiration, body temperature, orientation, and remembering, function unconsciously. In short, our bodily experience of time is a very rich one.

One of our many internal clocks literally ticks out its rhythm—the beating of our heart. These rhythms are in turn evoked in the drumming and dancing of people all over the world, from rituals in an African village to a rave in a London warehouse. But as we've discussed, this measure of process, this natural clock, has an inner fractal nature. The stamping feet of traditional dancers, the drumming of a jazz musician, and the beat given by an orchestra conductor, are never totally exact and mechanically metronomic. Computer analysis shows that, like healthy heartbeats, the rhythmic intervals in such music are always slightly irregular. It is this fractal fluctuation within regularity that brings the music alive. The heart that has locked itself into a limit cycle is on its way to heart failure, but the heart that is open and fluctuating with fractal variance is vibrant.

Seeing time as a measure of process in touch with its environment accords more directly with our experience than seeing time as the equal interval ticking of a mechanical clock. We begin to get a feel for the different "times" of process when looking at time-altered photography. In a slow-motion portrait of an athlete running, we see the fractal movements occurring throughout the

Many different processes and measures of time are present in this photograph.
Photo by Richard Halliburton

runner's body. In a several-day time-lapse sequence of clouds over a landscape and plants growing, we glimpse some of the hidden, long-range pulses going on in the area.

For Polynesian islanders, life stretches out into slow motion at sunrise and sunset. Over what, to us, is a relatively short space of time, the light changes and the sky moves through a spectrum of colors. For the islanders, sunrise and sunset are times when boats are put out to sea and fishing begins. Consequently, during this period a number of "hours" pass, some only a few tens of our minutes long. But in these island "hours," a great deal of activity takes place. In the middle of the day, when the hot Sun is high in the sky, people sleep or do the minimum of work. Then the "hour" is more than 100 of our minutes long. With our mechanical conditioning, we would say the islander's hours are of unequal lengths. They would argue, out of their long experience of living in an environment that changes rapidly two times a day, that each hour was of an identical length, for it contains the same amount of process.

The Polynesians have harmonized themselves to the flow of time in their environment, and if their variable hours seem a little odd to us, it is because we have become conditioned to synchronize ourselves with the mechanical clock time of our industrial environment. We're never totally successful in this synchronization, however, because our inner experiential time refuses to be pinned down to equal intervals. Our major problem comes with the common delusion that only the external time of the clock is real and so we'd better learn to fit in. It is somewhat ironic that a popular television game, appearing in different variations in much of the world, is called *Beat the Clock*—because the clock is the one thing in our modern world that we can't beat. In trying to beat it, we ourselves become mechanical. When we can't match up to the clock, we get jumpy, become stressed, and are split from our inner being.

David Shenk, author of *Data Smog: Surviving the Information Glut*, says that many of us are now so caught up in the pace of computers (faster versions now appearing every few months) that we are becoming like the person in the elevator who keeps pointlessly smacking one of the floor buttons to hurry up the ride. Shenk says, "In our compulsion to improve efficiency, we easily forget that intelligent work by humans is not just a matter of processing speed. Notice the constant stream of spelling mistakes and missing words in the E-mail you receive. Good work takes time and patience."[4]

The more we try to couple our internal flexibility to the external beat of a mechanical time (such as the processing speed of a computer), the more our internal fractal self-organization is threatened. In each of its lessons, chaos theory suggests that we connect, live within the center of complexity, and enrich our feedback loops with the world. Here, it suggests that we restore our bond with the rich time of nature and our own internal clocks.

Time drags its feet when we're bored, but a whole afternoon can flash by when we're engaged in something. In which situation do we have "more time"? The conventional mechanical model suggests that both times have equal duration. Yet when the afternoon flies by, we feel we don't have enough time. Trying to measure inner time using a clock creates confusion about how much time we have in any given situation. The fractal perspective, however, allows us to ask a different question. Which time has significance for us? Our boredom left one time empty; our passion and enthusiasm make another time rich and multifaceted. And so we don't need "more time," but time that is fuller—fuller not in the sense of getting a lot of things done, but in the sense of engaging the processes taking place.

The fact that time is process becomes evident when we meet people who have detached themselves from its movements. In Charles Dickens's *Great Expectations,* Miss Havisham, deserted on the eve of her wedding, cuts herself off from time and, dressed in her bridal clothes, lives out her life in a single frozen moment. Because her fate was too traumatic to face, Miss Havisham stopped the clock and refused to experience any more time. She felt she could only survive by living in the brief happy moment before the painful news was broken to her.

In William Faulkner's story "A Rose for Emily," the main character also attempts to freeze time, and Faulkner reveals that an effort to stop time actually has the opposite effect: It turns natural aging into putrefaction, isolation, and living death.

Miss Havisham and Emily are fictional characters, but we probably all know people who have tried to stop time in their life to some extent. We probably have encountered that type of forty-year-old who appears physically much younger and has all the mercurial energy and enthusiasm of a teenager. Psychologists refer to him as the "Puer Aeturnus," the eternal youth who fears the responsibility of maturity, avoiding decisions and commitment and shutting himself off from the processes of life. The Puer's whole life is waiting to be lived. He is constantly making plans but never making decisions. Isolating himself from time allows him to remain physically youthful, but at the cost of becoming elusive and evasive. The Puer restricts his outer connections to the world, and if his face remains unlined, it is by virtue of the chaos he is continually creating around him.

His counterpart is a man prematurely aged at twenty-five or thirty, serious, pessimistic, distanced, and dry. The Puer is mercurial while the so-called Senex type is identified with Saturn, the bringer of old age. The Puer abstracts himself from process in order to avoid time's aging. The Senex constantly anticipates the station at the end of the line.

The Senex and the Puer, as negative extremes of an unbalanced personality, suggest the extent to which the physical age of the mind and body is less about clock and calendar than the way we relate to the processes of life. We have also all met old people who are active and creative, not from an attempt to remain young but from retaining and constantly rediscovering their ability to enter the fullness of time.

Creative Time

A potter, now working in Ireland with the Japanese Raku method of pottery (a method developed for making vessels for the tea ceremony), describes his creative encounter with time: "People watch me taking a pot out of the furnace. It all happens very fast, but for me I'm there in each microsecond, and every moment is different. There is so much happening in that split second—touching the pot with the tongs, taking it out, exposing it to the oxygen in the air. An enormous amount of chemistry is going on and I have to be right there at every moment. If I'm not in the right frame of mind, the pot will be a failure. Even knowing that the temperature is right, you couldn't use a thermometer or a watch for that. It's almost something you can't see, because the furnace is so hot. It's just a little scintillation. You have to be there with it and feel the time inside yourself."[5]

Being in the moment means putting yourself at the swirling wall of the vortex where the movement between you and the not-you is taking place. Creative people think of these rich times as "moments of truth"—times when they experience what it is to be authentic.

When our sole measure of time is mechanical, we experience time as a shopping basket that has to be filled to be brim. We have a number of tasks to do over the weekend and we know that we won't find time to do them all. So we push ourselves, rush things and lose the flavor of life. An executive's day may begin with a working breakfast, a rushed lunch, and a business dinner in which problems of negotiation destroy any pleasure in the food.

Others work exceptionally hard yet always seem to find time for a leisurely dinner. They are connected to the food and careful about the time devoted to it. Dinner each evening is not something bolted down in front of the television set, but an act of renewal for the whole family. Each dish has its own time associated with it; the food is there to be savored, and part of its enjoyment consists in the expansive time of talking and arguing with friends over the dinner table.

Dwelling in time in this way allows us to discover the individual rhythms of the day. We enter into the time necessary for each task and therefore experience a multiplicity of times simultaneously. Our individual creativity demands that each activity flower in its own time. The Zen artist may spend hours, days, or even months meditating in front of a blank piece of paper and then finally render a butterfly alighting on a stalk of bamboo in a crescendo of gestures. We can ask, "How much time did the artist really spend making that drawing? A second? Perhaps months? Perhaps years? Perhaps a whole lifetime was needed before the gesture could be made."

Creative people, and we are all creative, need a great deal of time (as measured on the clock) in which they are simply "doing nothing." To the outside world, they appear to be daydreaming or simply fooling around. But inside, they are connecting to the time of the work, to its subtle rhythms and fractal structures. Actress Glenda Jackson referred to the time needed for a character to grow during rehearsals as "putting the bread in the oven." Her remark evokes the idea of matter sealed in the alchemical vessel and placed within the hot internal darkness of the furnace. A significant feature of alchemy, which many psychologists take as a metaphor for human internal development, is the necessary time taken for each of its various stages. The "Work," as the alchemic project is called, cannot be hurried, nor can it be slowed down. Just as with our own life experience, each stage demands its own time.

Creativity may therefore demand long periods of apparent inaction. But it can also pour forth with amazing rapidity so that a tremendous amount gets done. Psychologist Howard Gruber suggests that creative people often employ a "network of enterprises," engaging in a multiplicity of tasks that, although different, turn out to feed into each other. Charles Darwin kept notebooks on a wide range of subjects such as zoology and geology. Each subject had a separate existence for him, but all fed into each other and enabled him ultimately to solve the puzzle of evolution. A creative life requires giving attention to things in a way that allows each endeavor to grow in its own way from the nourishing context of all the other creative "enterprises" going on in that life.

So it is not so much that creative people work faster or harder than anyone else, or that they are able to cram a larger number of different activities into a single day. Rather, their many tasks are taking place simultaneously, each within its own time, and these times are coupled together, receiving energy one from the other. If we were to add up the totality of time that seems to be involved in a creative day onto a linear timetable, it would probably exceed a day's twenty-four hours. But creators make an alliance with the fractal dimensions of time, and time, in turn, gives them the time they need.

This same rich and expansive time is available to all of us, but our industrial society has conditioned us not to experience time in this way. If we do attempt several tasks or engage in a number of interests, we are accused of being a dilettante, unfocused, scattered, and flitting from one thing to another.

On the other hand, if we sit meditating at our office desk, we are accused of wasting time and so we'd better quickly find something to do. Bill, a physicist working for a research organization, once moved a big easy chair into his office. When asked what it was for, he said that he liked to sit and daydream, maybe even doze a little in the afternoon. The director was horrified, "You're not paid to sleep; you're supposed to be working all the time you're here." It didn't help when Bill pointed out that he was publishing far more than his colleagues, but he needed time to daydream in order to come up with new ideas. Dreaming for Bill was entering into the fullness of time; to the bureaucrat, it was simply wasting the time the organization was paying for.

And so many of us seem able to do only one thing and arrive back home from that exhausted. If we do want to paint or write our memoirs, it's something we'll leave for a weekend or for our retirement, when we hope to have more time. But the truth is this: The time that we really want is the fractal time we have right now.

LESSON 7

Rejoining the Whole

Lesson about the Tide of a New Perception

Think back to the first time you saw that breathtaking photograph of Earth viewed from space. For most of us, the sight of that intense blue sphere, veiled in swirling clouds and inlaid with the fractal fretwork of continents, islands, deserts, rivers, mountain ranges, lakes, and polar ice caps, stirred in our depths something mysterious, moving, and even spiritual.

The astronaut Edgar Mitchell described his view of Earth as "a glimpse of divinity." He was profoundly moved by "this blue-and-white planet floating there, and knowing it was orbiting the Sun, seeing that Sun, seeing it set in the background of the very deep black and velvety cosmos, seeing—rather, knowing for sure—that there was a purposefulness of cosmos—that it was beyond man's rational ability to understand, that suddenly there was a non-rational way of understanding that had been beyond my previous experience." He recalled that, on the trip home from the Moon, "gazing through 240,000 miles of space toward the stars and the

Photo by NASA

planet from which I had come, I suddenly experienced the Universe as intelligent, loving, harmonious."[1]

A Russian cosmonaut, Alekesi Leonov, reacted to the whole Earth he saw from his space ship as "our home that must be defended like a holy relic."[2]

Lewis Thomas was inspired by these photographs to compare the Earth to a single human cell. A cell is a fluid-filled membrane containing mitochondria, centrioles, basal bodies, and "a good many other more obscure tiny beings at work," each with its own history and separate evolution. Yet in a miraculous way all come together to form a completely interdependent and unified entity.

A single cell, Thomas argued, is a fractal microcosm of what life on Earth has achieved.

But there are also ironies about this image from space. For example, why did it need all the technology of our modern industrial society, including an intense competition for dominance in space, to bring us to a place where we could see the indivisibility of life—a vision that each one of us could recognize, because we already knew it in our hearts?

But again, the irony: Beneath this glimpse of life's wholeness lie national boundaries, property lines, busy roads, sectarian and racial strife, competing interests, accelerating conflicts, and our competing selves. The Earth humans have redefined over the last few hundred years—an Earth where human activity has shriveled the planet's protective layer of ozone, greedily hacked down the rain forests, and genocidally eliminated thousands of species—is the antithesis of that fluid, integrated, stunningly beautiful "cell" our representatives gazed down upon from space.

What's important about this image of our blue planet is the shift in perception that goes deeper than a mere change of viewing point. It's the subtle mental shift, the reorganization of the entire way in which we conceive of our world.

The writer and physicist Fritjof Capra believes that the human race is experiencing such a "crisis of perception." The fragmented, analytical view of reality we have lived with for so long is, Capra argues, "inadequate for dealing with our overpopulated, interconnected world."[3] Our worldview is the medium in which we mentally swim. It's so much a part of our surroundings that we take it for granted and don't notice its pervasive presence. But seeing Earth from space draws our attention to it, because the image recalls us to a worldview dramatically different from the one we've been immersed in for so long.

Chaos theory, like the image of our incredible planet in space, offers us a perception and an associated conception of an interconnected world—a world organic, seamless, fluid: whole.

Wholeness is the central theme of mystical revelations the world

Religions throughout the world insist the cosmos is an indivisible unity. A mirror of the cosmic wholeness, this ancient Hindu temple, with its many gods and figures, was carved out of a single huge stone. *Photo by John Briggs*

over. The Hindus seek unity of the individual soul or "Atman" with "Brahman" or "World Spirit," or All. For Christian mystics like the twelfth-century St. Bernard of Clairvaux and the fourteenth-century St. Catherine of Siena, the totality of God's love overcame all human contradictions. Among many traditional peoples, wholeness is a way of daily life. When Native Americans say "all my relations" during a ceremony, they are expressing their relationship not only to members of their group but to the plants and animals; rocks, trees, and rivers; the dwelling place of the sky and the soul of the Earth; their ancestors and descendants; and the multitude of energies that power the cosmos.

The ancient Chinese *I Ching* is based on a holistic cosmology in which the relationships between Heaven and Earth, mountains, fire, wind, and wood are reflected in the state, family, and lives of individuals. In a similar vein, the medieval alchemist distilling the

The well-known Chinese *tai chi* circle depicts the fluid interplay between the movement of light and dark, heaven and earth, active and passive, manifest and nonmanifest, being and nonbeing, making it a picture of the ceaseless, undefinable, holistic flux.

philosopher's stone in his laboratory was said to be emulating the whole by engaging in a primordial act like the one that created the Universe. The *I Ching* and the theory of alchemy are examples of the perennial philosophy that speaks of a self-similar mirroring of the cosmos within each of its parts.

The Old Perception Shift: From Medieval Holism to the Rise of Mechanism

Although chaos theory returns us to an ancient understanding that the Universe is whole, it also brings very new insights to bear on this idea. These new insights have emerged in part from the fact that the new perspective of wholeness is being born out of a mechanistic perspective that is the *antithesis* of wholeness. This mechanistic perspective is the one we have known for the last several hundred years. In its day, that perspective was born from another kind of holism that existed in the Middle Ages.

The cultural perception shifts we're referring to here were mind-shaking events. They tremendously transformed the way people thought and behaved. Looking at these shifts in a little detail will perhaps help us glimpse the kinds of revolutionary effects that could take place if we fully engaged the radically new holistic perception that chaos offers us.

During the medieval period, from about 600 C.E. to about 1400 C.E., a certain kind of understanding about wholeness prevailed throughout Europe. The Earth was considered a living being, and the human artisan was an assistant or midwife to nature. Metals grew in the womb of the Earth. The miner, smelter, metalworker, and goldsmith were engaged in the sacred tasks of helping nature reach perfection. Medieval astrologers adhered to the doctrine that the character and destiny of life on Earth is unified with the movements of the stars in the sky, "As above, so below."

When Abbot Suger ordered the rebuilding of the Abbey of St. Denis in the eleventh century, he did not regard beauty as something merely pleasing to the eye, but as an expression of essential goodness and truth. For Suger, God was immanent in the natural world; He could be found in forms, colors, and, above all, the light that flooded into the abbey. The abbey was a microcosm of Earth and the cosmos.[4] The words "goodness," "truth," and "God" were all rooted in the idea of a unified, single cosmos. Dante's *Divine Comedy* is a grand poetic metaphor for this divine holistic order. His circles within circles provide an image of heaven, the astronomical cosmos, human society, the medieval walled city, and the levels of human consciousness and spiritual development.

The seeds of the scientific and mechanical perspective in which we have been living began to take shape around 800 years ago as the European psyche moved to differentiate itself from the rest of life on Earth. Little by little over the next centuries, nature became objectified and externalized, and with it grew the idea of humans as fundamentally separate individuals with their own aspirations and inner life.

This change is even reflected in the word "consciousness." We understand our consciousness as the essence of our individuality. However, until the Renaissance in the fifteenth century, consciousness was not considered to be the exclusive property of individuals. The Latin roots of the word are "con," or "with," and "science," or "knowledge." Before the Renaissance, "consciousness" referred to what people knew together, not what they were aware of as individuals.[5]

Historians point to a variety of seeds for the Renaissance transformation. In theology, Thomas Aquinas denied the view that humans participate in the intimate workings of nature. He argued that in goldsmithing and other crafts, the essence of matter is untouched and only its external form is modified by the labor of hands. That implied that matter is indifferent to our actions and desires, and therefore nature must be external to us.[6]

Another seed was the invention of the printing press, which brought new learning to the literate and encouraged private, individual study. The rise of city-states fostered the birth of a new type of authority: rulers who gained their position through personal ability, influence, and charisma rather than through hereditary power. Being an individual now became a virtue. By the time of the Reformation, human reason was challenging Divine Revelation and the traditional authority of the church.

A profound metamorphosis of consciousness slowly but inexorably seeped into the medieval conception of reality. The change was probably so subtle that most people were unaware that they were in effect evolving into a new kind of human.

By the time of the Renaissance, "Man" had become the measure of all things. Before the age of Shakespeare, characters in drama existed externally in the form of the personas of the Italian *commedia dell'arte*, or as manifestations of the four humors or temperaments (choleric, phlegmatic, melancholic, and sanguine). It was Elizabethan genius not only to understand character but to reveal, through the device of soliloquy, the way a Hamlet or a Macbeth grappled with their inner contradictions and motivations. Where lesser Elizabethan playwrights used soliloquy to further plot and supply information, Shakespeare made it the arena of personal psychology in a way that would have been more than inconceivable a few centuries earlier—it would have been incomprehensible.

In art, Leonardo Da Vinci's paintings portrayed real human figures rather than symbols of a drama of cosmic unity. In doing so, he risked the death penalty for blasphemy when he performed dissections on dead bodies.

Before the Renaissance, artists were to a great extent anonymous, not far from artisans and craftspeople in status and eager to serve religious ideals. The artisan who sculpted an ornamental device onto the high arch of a Gothic cathedral knew that no one else would see it. Yet God would witness the work, for it was an expression of God's will. By contrast, the Renaissance created the myth of artist as hero and genius, an identity, the individual expressing his individuality, mastering his materials. Art flowered, but there was a price. The more individuals separated themselves from society and the natural world, the more that world became distanced and objectified for them. It is no accident that Renaissance art is primarily associated with the development of perspective, a geometrical technique in which the world is literally projected outward and seen at a distance, as if through a window. Certainly other cultures had known the trick of one-point perspective but had chosen not to exploit it, because to make the illusion work, it is necessary to distort the shapes and forms of objects until they all conform to that one obsessive viewpoint. This meant that the individual painter's point of view was replacing the encompassing, omniscient perspective of God.

The worldview that began in the Renaissance continued to proliferate over the following centuries. In the early seventeenth century, a concerto meant a group of instruments playing harmoniously together, but by the middle of the eighteenth, it was the struggle of a single instrument, a piano or violin, to assert its individual vision against the whole weight of an orchestra. Literature saw the development of the novel and the biography. By the time of Beethoven, lone individuals were shaking their fist at both God and Fate.

The Renaissance's increasing emphasis on the separateness of individual human consciousness led to a conception of nature as a vast collection of objects that could be subject to scientific investigation and experiment. In the seventeenth century, Isaac Newton stabilized the rising structure of the scientific enterprise by generalizing the observations made by Galileo and others of the motions of falling bodies, swinging pendulums, and planetary orbits into three laws that would describe the working of the entire cosmos.

Sixty years earlier, British philosopher Francis Bacon had asserted that "knowledge is power" and that such knowledge could be gained by putting nature on the rack to extract her secrets. Now Newton's equations completed the dehumanization of the natural world by picturing it as composed of mechanical building blocks in interaction. Understanding became a question of breaking things down into their components and explaining the causal links between them. Nature became a great clock that science could take apart and reassemble, and this became the overriding metaphor of the scientific enterprise. Prediction and control were the driving forces of a new scientific society.

Control had been the province of governments. The Newtonian dream perfectly complemented that ethos. Science's associated technology amplified the power of control through its ability to channel enormous energies, develop new substances, transport material at higher and higher speeds, and weave a network of communications around the Earth.

Science and society fed back into each other, expanding the scientific worldview enormously. Today, society provides the resources needed to build highly expensive particle accelerators, fusion chambers, space telescopes, and the like. Science, in return, supplies an endless stream of new technological objects—from land mines to cellular phones and synthetic food—as well as a flow of new ideas that reinforce the societal and scientific goals of prediction and control. In its very success, science has intensified the mechanization of our society and confirmed our perception of a mechanical universe.

The extent to which this mechanistic worldview has become the medium in which we are submerged can be immediately grasped by listening to a call-in radio show featuring talk therapy.

Typically, a caller explains in a minute or so some psychological dilemma involving a mate or "other," and the psychologist propounds a thumbnail analysis and a course of action. Often the caller is advised to seek professional help to repair some evident damage or deformation to the self. For example, the self needs to become more assertive or more sensitive or less phobic. There may even be a series of steps laid out in order to accomplish this self-reconstruction. On one television talk show, a bewildered husband who had lost his job and discovered that he actually enjoyed staying at home to do housework turned to the expert psychologist and asked plaintively, "How should I feel about this?"

So, as the twentieth century draws to a close, we have also encountered the dark side of that path we began to tread 800 years ago when we separated ourselves from nature. It is certainly true that this path led us to the glittering flourish of art, poetry, music, architecture of the Renaissance, and on to the scientific and technological developments that followed. But it also brought us to the wasteland of progress and unlimited growth that are now so much part and parcel of the mechanistic paradigm. Both concepts are inherently flawed. Unlimited growth obviously can't be sustained indefinitely. At some point the use of planetary resources will outrun our technological ingenuity. Historians of technology have shown that what advertisers vaunt as progress is often little more than a fad or fashion that substitutes one product for a newer one. Again and again, we see how new "advances" bring with them unforeseen side effects. Indeed, it becomes increasingly difficult to tell if all our "progress" is in fact leading to the general improvement in the quality of life.

The mechanistic paradigm continues to bring with it profound moral problems. Biologist Brian Goodwin warns about the way the mechanical worldview is playing out through genetics: "According to the current biology, genes determine organisms, and organisms are simply accidental collections of genes that are functionally useful to us human beings. Therefore, it is perfectly legitimate to change the genetic composition of an organism to suit our needs. We can create chickens or turkeys with enormous amounts of breast meat, even though these animals cannot reproduce and cannot live a normal life. It's okay to change them this way.

"But such things are deeply wounding to our relationship with the natural world and with each other because it means turning everything in life into a commodity. It encourages me to think of you as just a bunch of cells or genes. These all have potential commercial value and to me, that's suicide. Organisms are not merely survival machines. They assume intrinsic value, having worth in and of themselves, like works of art."[7]

In making criticisms against mechanistic science and technology, we must be careful not to dismiss out of hand the benefits it has brought us. Mechanical science has helped us live generally healthier and longer lives than our counterparts in the Middle Ages and to experience the world in marvelous new ways. But it does seem fair to say that our nearly total immersion in the mechanical, reductionist approach has led us as a society to forget our instinctive empathy for the natural world. And it has produced a way of thinking in which we tend to treat ourselves and others as objects for manipulation. The American philosopher Henry David Thoreau had already seen the dilemma when he wrote: "Lo! Men have become tools of their tools."

The New Perception Shift: From the Mechanical View to Chaotic Wholeness

The mechanistic worldview took several centuries to flower from the original seeds sown in the late Middle Ages into the present triumphs of science and technology. At the turn of our own century quite a different seed was planted, this one by the French physicist and philosopher Jules-Henri Poincaré. Today we are witnessing its unfolding in chaos theory. Significantly, when Poincaré caught the first glimpse of chaos, it was not in terms of a disorder and lawlessness in the Universe. What he saw was that chaos is wholeness.

Poincaré planted this germ of chaotic holism at the end of the nineteenth century, a moment when technological optimism and faith in a mechanistic worldview were at a high-water mark. The era marked such marvels as the Eiffel Tower, the automobile, the first experiments in radio transmission, and Nicola Tesla's electrical generators that harnessed Niagara Falls to light the city of Buffalo.

Poincaré was not seeking to overthrow this mechanistic program, but to extend it and make it even more comprehensive.

From the time of its inception, the pristine Newtonian universe contained the troublesome blemish that its mathematical approach was only capable of describing, in an exact manner, the motion of two mutually interacting bodies and not three or more. Physics was eminently capable of working out how each planet orbits individually around the Sun. But the solar system obviously doesn't consist of a single planet; it contains many planets, moons, planetary rings, and a whole belt of asteroids. Although the solar orbit of each can be computed for tens of thousands of years to come, astronomers didn't know how to take into account such fine details as the tiny deflection on the orbit of an asteroid caused by the pull of the planet Jupiter. In their search for greater predictability and an all-encompassing description of nature, scientists needed to solve this so-called three-body problem.

The accepted, though unsatisfactory solution of the day involved "approximations." Because the pull of Jupiter on an asteroid is very small, astronomers could make a series of estimates as to effects and then add them all up. In many other real-life situations, scientists are forced into such approximations (Perturbation Theory as it is generally called), and they seem to work quite well. But the more philosophical physicists were uneasy about fudging planetary orbits that would exist for thousands of millions of years.

Poincaré set out to tackle the three-body problem head on. As he worked at the complicated mathematics, he discovered that under most conditions events were as astronomers and physicists expected—the weak gravitational pull of a second planet on a planet or asteroid orbiting the Sun has an almost negligible influence. But he also found that under certain critical conditions, something quite extraordinary happens. The tiny corrections begin to accumulate, feeding back into each other, until their overall effect on an asteroid's orbit causes it to wobble, swing violently and erratically in its orbit, or even fly out of the solar system all together.

Poincaré had stumbled upon chaos. But he had also stumbled upon a significant paradox. This chaos only exists within the solar system because the entire system is holistic. Though chaos appears to be the opposite of wholeness, Poincaré realized that wholeness lay at its heart.

If the planets were relatively independent of each other, then it would be perfectly valid to consider the effects of a third body as a simple perturbation. However, because of the nonlinear effects of feedback, planets cannot be treated as if they are essentially independent, as they were in the mechanical perspective. To take a specific example, Jupiter's pull on an asteroid's orbit around the Sun is very small. Earlier, astronomers had assumed that this tiny attraction would only shift the asteroid's orbit by a small amount. But as the asteroid's orbit shifts, so, too, does the strength of its mutual attraction to the Sun. This produces another movement in its orbit. In turn, these orbital shifts produce small changes in the gravitational force the asteroid experiences as it nears Jupiter.

Normally, all these tiny changes end up producing only a minor correction to the asteroid's solar path. Yet under certain critical conditions, the various shifts and changes in orbit and gravitational attraction act to feed back one into the other, accumulating until a *resonance* occurs and the whole effect blows up into chaos.

Resonance happens when systems vibrate or swing in sympathy with each other so that the tiniest connection between them progressively magnifies their mutual interaction. Place an E tuning fork on a badly tuned violin and nothing will happen. But if the top string is in tune, it will begin to resonate in sympathy with the fork. If you want to go higher and higher on a playground swing, then you time the pumping of your legs to coincide exactly with the top of each swing. When systems work in sympathy, big changes can result from the cumulative effect of tiny interactions. We recognize this now as the butterfly effect discovered in the early 1960s by Edward Lorenz. What Poincaré had discovered within the "Newton's clock" of the solar system was that under the right circumstances, it is possible for Jupiter and an asteroid to go into resonance as they orbit around the Sun. The feedback loops that connect them loop 'round and 'round—like the screech the speaker makes when the microphone has been placed too near it—causing the asteroid's orbit to be chaotic.

For more than half a century, Poincaré's result remained like a theoretical time bomb in an otherwise orderly and mechanistic paradigm. In the years before modern high-speed computers and new techniques in mathematics, its implications simply could not be explored. But by the 1970s, as ideas of chaos theory began to proliferate, blank regions were actually discovered in the asteroid belt. Similar empty regions had already been observed in the rings of Saturn. Scientists turned back to Poincaré's work and realized these blanks are the places where the chaotic orbits he predicted would exist. An asteroid or other of lump of space matter that tried to inhabit such a region would become trapped in a net of feedback loops, its orbit growing progressively wilder, until it finally flew off into deep space. In retrospect, chaos theory scientists understood what Poincaré had glimpsed. Because the solar system is holistic—with planets, moons, asteroids, and comets constantly feeding back into each others' orbits—some regions become chaotic zones. These are proof that within the cosmos, chaos and wholeness are entwined.

Poincaré's discovery illustrates the difference between chaotic wholeness and the symbolic wholeness of the medieval alchemist, the dual (yin-yang) wholeness of the ancient Chinese or even the wholeness of the Romantics who sought the experience of an enveloping nature where all particular things seemed to vanish. Different from these, chaotic wholeness is full of particulars, active and interactive, animated by nonlinear feedback and capable of producing everything from self-organized systems to fractal self-similarity to unpredictable chaotic disorder. In what is perhaps a joke by the cosmic trickster, chaotic wholeness now celebrates the very phenomena that were dismissed as "messy" and "accidental" in the mechanical paradigm.

The image of our blue Earth from space is an icon to this new holistic perspective. We can now see that from the fractal shape of the planet's continents and the flow of weather patterns, right down to individual living cells, all of it is an enfolding of self-organized chaotic systems within systems.

Zoom in from the satellite photo of the globe to the rain forests of Amazonia. Huge areas are being destroyed. Why should we care? The mechanistic answer is that if things truly get out of hand, we will still be able to control the situation by replanting the trees and managing the forest. But chaos theory tells us that our interventions are limited and that their outcome is always, to a certain crucial degree, unpredictable. Assuming that future gener-

A photo of an interconnected chaotic system driven by feedback. *Photo by John Briggs*

ations will be able to develop the technology to clean up the mess we are making now is a dangerous delusion.

Forests, particularly rain forests, create their own climate. They retain moisture and bind even the thinnest topsoil. Within their canopy, microclimates foster a huge variety of plants and animals. Cut down the forest and you also destroy a complex network of feedback loops and irrevocably dismantle an essential part of the Earth.

To see what may happen to our future generations, zoom in on the bleak English Pennines of Emily Brontë's *Wuthering Heights* or the bare majestic mountain peaks of Wordsworth's Lake District and ponder why over the last two thousand years no trees have grown in a location that was once heavily forested.

Thousands of years ago, Neolithic farming was responsible for the destruction of the temperate rain forests that once covered much of Britain. Producing microchanges in the local climate, the ancient farmers broke the loop in the circulation of water by transpiration. Topsoil was washed away by rain and vegetation rotted to create acidic soil. The result was an irreversible change, a region now covered with bogs and sparse grass. Something like that could be the fate of the Amazon basin and the Russian taiga, which lumber companies are beginning to dismantle. Scientists believe that such massive destruction of the world's forests could seriously alter the Earth's climate.

Won't the trees just grow back? Lakes, forests, and rivers have always been prone to natural disasters. Sometimes they are wiped out by these events, but in many cases their rich web of feedback loops makes them so flexible and adaptable that they adjust and weather even a dramatic change. But we must be on our guard, remembering the nonlinear behavior inherent in chaotic systems. Forests may be marvelously resistant, but once we stress them past a certain point or change their environment too quickly, they may jump abruptly into new behavior or even collapse.

This warning is present for all of us to read. The natural processes on Earth are indivisible, constituting a holism that must be nurtured and maintained. Push the system too far and it can break down. Lewis Thomas imagined all of Earth as a single cell. Scientist James Lovelock has explored the idea of a holistic Earth in which organic and inorganic systems are interlocked together in a way that can be envisioned as a single living being he calls Gaia, after the ancient Greek goddess of Earth.

Lovelock came to the idea of Gaia after studying the composition of gases within the Earth's atmosphere. To take one example, methane (natural gas) burns in the presence of oxygen to produce water and carbon dioxide. Lovelock realized that something curious must be going on when a planet maintains considerable quantities of methane within an atmosphere rich in oxygen. By rights, all the free methane and oxygen should long ago have reacted and burned up.

Both methane and oxygen are, of course, by-products of life, which is constantly producing them. Lovelock drew a significant realization from this fact. Methane, oxygen, sulfur, ammonia, and methyl chloride are all present in our atmosphere at quite different concentrations from what would be expected in the inert equilibrium state, that random state that would prevail if the existing supplies of chemicals were allowed to mix together in an atmospheric beaker. This is also true of the constant percentage of salt in the sea—despite the fact that millions of tons of salt are annually washed into the world's oceans from rocks and soil. What Lovelock found so striking was that these concentrations also happen to provide the optimum conditions for the continued survival of life on Earth.

Lovelock's dramatic deduction was that life as a whole is carefully regulating the planet. The entire planet has evolved as a living being, with the forests and oceans as its organs.

Within Gaia, the simplest of organisms, bacteria and plankton, play the most important roles in maintaining the environment that allows more complex organisms to exist. In turn, these complex organisms, from termites to cows, supply bacteria with what they need. The Earth, as Gaia, is a complex coevolving entity comprising microorganisms, grasses, trees, animals, climate, and even the movement of continents. The constant activity of feedback at all levels maintains the dynamical and far from equilibrium structure of the dazzling being we see from space. This is the reason why life is so resilient to accidental damage and changing conditions. But that chaotic nonlinearity also means that the Earth is delicate—vulnerable to the impact of the unrestrained human technology fashioned by the mechanistic perspective.[8]

Chaotic self-similarity echoes down from the planet to the individual cells of our bodies. Each one of us is a set of dynamical relationships among entities we cannot really be said to "own." As Lewis Thomas has said, the mitochondria that are found in the interior of each cell of our body possess their own DNA, which is quite separate from ours. In fact, they are the descendants of bacteria that entered the ancestral precursors of our first cells in an act of symbiosis and mutual support. This interlocking cooperation does not stop at mitochondria, but extends to many more organisms, which together make up the ecology of our bodies, including the spirochete bacteria that became our brain cells.[9]

Chaotic Wholeness and a Different Approach to Life

When an automobile breaks down on the highway, we open the hood and look at the engine for a defective part. That approach works perfectly well, and we'd have to be more than a little idealistic to think that a broken fan belt or blocked fuel line was the result of the car's lost vision of wholeness. But families, societies, and ecologies are not machines. Chaos theory teaches us that we are always a part of the problem and that particular tensions and dislocations always unfold from the entire system rather than from some defective "part." Envisioning an issue as a purely mechanical problem to be solved may bring temporary relief of symptoms, but chaos suggests that in the long run it could be more effective to look at the overall context in which a particular problem manifests itself.

Recently, the mental illness of manic depression has received considerable attention. Millions are now thought to be afflicted with some form of this disease, which is usually treated with medications like Prozac. But if manic-depressive illness is on the rise in our population, shouldn't we be looking closely at the society in which this epidemic occurs?

Our traditional mind-set has focused on social, political, and ecological problems as lying outside ourselves. As a result, we try to overcome problems by conquest or negotiation, which has the effect of reinforcing our perception of inherent separation.[10] From deep within this mind-set springs the violence that today dominates much of our consciousness. Look at the language we use to describe society's problems. We declare war on poverty and addiction. Doctors use "aggressive" methods on the critical patient, drugs are described as magic "bullets," and we are given "shots" to "fight" disease.

There is a slowly growing recognition that diseases like cancer may not have a single cause that can be knocked out by magic bullets. Each cancer appears to be the result of a host of "cofactors" existing in a unique combination of feedback for an individual: low-level radiation and chemical exposure from the environment (an increasing factor for all of us), diet, lifestyle, genetic background, exposure to prior diseases, psychological stress, significant interpersonal relationships. All of these interact. The "cure" for cancer, or its successful treatment, may be more dependent on addressing the whole person's life than on any magic medical bullet.

Medical problems, societal problems, and individual problems all have a holistic dynamic. So should we declare war on drugs or begin to seriously inquire into the interlocking factors within our society that cause so many young people to turn to drugs in the first place? Do we support massive funds for dawn raids on narcobarons or do we make the connection that international agreements to hold down coffee prices make poppies a more profitable crop than coffee plants? In other words, instead of projecting the problem, shouldn't we focus on how drug abuse is related to who we are as a society in the modern world? From that holistic focus, perhaps a new kind of action could emerge.

A mechanical perspective that sees the world and ourselves as no more than a collection of externally related parts blocks the clarity of our vision. Lovelock points out that if you only examined individual cells in the body and not their overall feedback interaction, you would never be able to guess that the body as a whole has the capacity to regulate the temperature of its entire system of cells. Likewise, we don't know at this point what it would mean for the creative capacity of human consciousness to work as a whole across the planet. Since the time of the Renaissance, we have focused on our existence as isolated individuals rather than as aspects of a "con-science"-ness—what we are in our knowing together.

Is it possible that we can shift our own perception to embrace the self-organized and chaotic whole? This idea may seem mysterious at first, yet an understanding of wholeness is already woven deep within us. There are moments in everyone's life of cooperation and spontaneous organization.

In 1993, when Hurricane Andrew raced across southern Florida and devastated the landscape, people came from all over the United States to help. No one really organized this; it was a simple act of compassionate self-organization. Steve Rodriguez rushed to Dade County from Waco, Texas. "I want to help," he told a *National Geographic* writer, his voice quavering with urgency. "I saw it on the news. I couldn't bear it. I told my boss and my wife I got to go there. I'm a certified forklift operator. I know CPR. Who can I talk to?" Joy McKenzie of Jacksonville drove down to wash people's hair at a church relief center. "My heart broke," she said. "I had to do something. I'm a beautician, so this was it." A contractor and a team of workers loaded up U-Hauls with food and supplies to rebuild a gym at the Miccosukee Indian reservation, hit hard by the storm. Asked why he was doing it, the contractor said, "Well, we're a bunch of queens who want to help. Every one of us. My lover has AIDS. AIDS is a human problem. So is this. We care."[11] Carpenters from New England packed their trucks with plywood and drove for two days to help the communities rebuild. Farmers in the Midwest sent milk and produce. Many of these helpers weren't part of any aid organization, they were individuals who never thought to get a reward for their efforts. Somehow they felt called by their deep sense of connection to do something. Their appearance on the storm-struck scene was an expression of their interconnection with the whole.

One Native American tells the story of how, as a young man, he used to travel across the United States and Canada attending powwows. He had little money but found there was always someone to give him a lift to the next reserve or offer him a meal and a bed for the night. He never needed to plan his itinerary or worry about bank machines and credit cards. He simply trusted the system and the old saying "All my relations." Now, in middle age, he has a demanding job but still finds time to visit the powwows. When he sees a young person hanging around, he offers him a lift or buys him a meal. For this man, the web of connections is strong and survives from generation to generation.

Underneath our feelings of isolation and our loneliness as separate individuals vibrates a sense of belonging and interconnection to the whole world. One curious piece of evidence for this is the "guilt" psychologists notice in the survivors of an air disaster or mass hostage taking. When people are killed around them, even though they are complete strangers, those left behind experience anguishing remorse that they did not die in the place of others.

Survival guilt suggests that at the basis of our psyche lies a sense of solidarity with the entire human race. The psychotherapist Viktor Frankl reports how he discovered this solidarity as a young man in the most unlikely of places, a Nazi concentration camp. Trapped in one of humankind's worst perversions of the mechanistic ideology, he was stripped of his identity and reduced to a number tattooed on his arm.[12] Frankl knew that he could easily succumb to seeing himself as a meaningless object, readily expendable fuel for a terrible machine.

One day, while he and others were being herded along an icy road to a slave labor site, he began to think about his wife "with uncanny acuteness." He didn't know whether she was alive or dead but suddenly realized that it didn't matter, because he was connected to her by a love that went far beyond her physical person. In fact, this love for her was connecting him to a holistic insight, an authentic truth: "that love is the ultimate and the highest goal to which man can aspire."[13]

Frankl was able to survive the concentration camps with his humanity intact. But was his insight about unifying love merely a sentimental idea, the psychological defense mechanism of a man at the extremes of suffering? This is quite probably how it would be viewed from our current scientistic perspective.

From the vision of the chaotic whole, we recognize that Frankl was able to move, inspired by what he experienced as unconditional love, from a perception of himself as an isolated victim to a profound sense of meaning and connection with the world outside him; this seems something like what Conrad implied when he described "solidarity" as the knitting together of the loneliness of innumerable hearts and "contacting the sense of mystery surrounding our lives."[14] Frankl's experience illustrates that an encounter with the terrible unknown of chaos can bring with it the apparently paradoxical feeling of an intimate, transcending faith or trust in a nurturing cosmos.

Experiencing solidarity with the whole universe is about freeing ourselves from the chronic habit of thinking that we're just disconnected fragments. It's about moving from an emphasis on the isolated self, from the consciousness of what we only know individually, to the consciousness of what we also know together. It is about moving from the old focus on individual heroic competition against the world to coevolution and collaboration. It is about moving from seeing nature as a collection of isolated objects to experiencing that we are an essential aspect of nature's organization. It is realizing that the observer must always be a part of what he or she observes. It is about moving from an exclusive emphasis on logic, analysis, and objectivity to an ability to reason aesthetically in a way that includes analysis but acknowledges its limits. It is about moving from obsessive focus on control and prediction to a sensitivity toward emergence and change. It is about a new understanding of time and our path in it. It is about using our subtle influence to become the participators of the blue planet rather than its managers.

As we enter into this new perception, we needn't entirely reject our earlier post-Renaissance understandings of ourselves as individuals and all the knowledge and technological advances that went with it. But in the light of chaos, each individual and collection of individuals may take on brand-new meaning as metaphors and fractals through which the whole is expressed.

After Words

Lesson 7.1325 . . . Missing Information and The (W)hole at the Center

A parable: Two friends decide to write a book on how chaos theory might apply in daily life. They divide the book into seven chapters, each explaining part of the theory and exploring its implications. For the next few months, they work away, busily sending each other drafts and notes until they realize they have left something out. Chaos theory is about being unable to predict and control. It's about never being able to make a complete description. It's about something scientists have called "the missing information!"

So where will they put the missing information? There's not enough room in the universe, let alone in a book, for everything missing. How on Earth will they tell people that once they've read this book the most important things will be the bits left out?

Then our friends remember the irrational numbers, one of the emblems of chaos. Irrational numbers find their own spaces, even when the number line is totally full. "We'd better make an irrational chapter," they decide, so here it is:

Photo by John Briggs

WARNING!!! DO NOT READ THESE WORDS.

• • •

A monk asked Baso, "What is Buddha?"
Baso answered, "This mind is Buddha."
A monk asked Baso, "What is Buddha?"
Baso answered, "This mind is not Buddha."

• • •

William McDougal, an American psychiatrist, asked his Chinese teacher to explain the nature of the Tao. At the end, he was still as confused as before and asked again. His teacher took him onto the balcony and asked him what he could see.

"I see a street and houses and people walking. . . ."

"What more?"

"There is a hill."

"What more?"

"Trees."

"What more?"

"The wind is blowing."

Finally, the teacher threw up his arms saying, "That is Tao."[1]

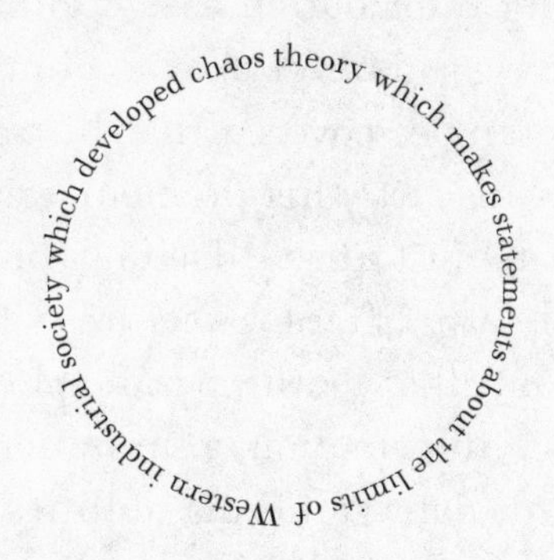

1. Every statement in this book is limited.
2. 1 is a statement in this book.

Paradoxes and koans take us to the edges of logical, rational, ordered thought. They cause our minds to run in loops and perform iterations of logic as we try to find a way out of the problem. Yet there can be no resolution from within the context in which they are framed. Koans tell us something is missing, something is incomplete about our concept of reality. Yet the very fact we think up such paradoxes in the first place means we are bigger than the conceptual systems we create. They tell us that we are the missing information we've been seeking.

Koans confront our desire to partition the world into dualities, to place concepts into convenient categories and then draw boundaries around them. By taking us to the edge of such thinking, they create the mental chaos necessary for creativity in which the mind shifts and self-reorganizes its perception of reality.

You can never round off an irrational number without leaving something out. The something you have left out is a hole in your information. At the same time, those dots at the end of the irrational number—the dots in the title of this chapter—are like stepping stones on a trail that leads to the whole of the system, to the hidden feedback loops, to all the little butterflies out there. Those dots of missing information are, in other words, a symbol for the whole, which we can never take account of...

When Edward Lorenz discovered that his second weather prediction didn't match his first, his problem was missing information. Rounding off to just three decimal places instead of six resulted in a completely different picture. Chaos theorists have been quick to point out that, both in principle and practice, there will always be missing information, a limitation to our knowledge, a hole in the data. Our data-gathering abilities can never be sufficiently extensive to know all there is to know about a complex system like the weather, let alone the world. For one thing, in a complex system there is no clear division between one "part" of it and another, which alone makes getting "all" the information impossible. For another thing, our act of trying to obtain the information, just our presence, perturbs a system in unpredictable ways.

Another koan. If chaos theory tells us about the missing information, isn't it also telling us that it is not the whole story? Chaos theory is science and all science is subject to change. In fifty years, time, the theory will probably look very different from the way it does today. But does that mean that the issue of missing information might somehow be resolved? Possibly, but very unlikely. Paradox and limitation appear inherent to our human thinking and existence. Whether chaos theory changes or disappears, it seems pretty certain that there'll always be a (w)hole of some important sort at the center of all our ideas.

Koan: You can't put the whole in your pocket and the reason is that your pocket is part of the whole. Therefore it has a hole in it.

In prior lessons, we've seen how the dimensions of space and time fractalize and look very different when viewed through the prism of chaos. Now we hold chaos itself up to reality and find an overlooked third dimensions revealed there: mystery. In fact, chaos's most timeless lesson may be that it reenchants us with mystery. It reminds us that amid the glitter and excitement of our expanding scientific knowledge we had forgotten about the unknowable that beats inside of all we know... Chaos makes a link to the experience that appears in the gaps of koans, poetic ironies, metaphors, and other self-similar amd -different forms... and to the feeling that appeared in our hearts when we first saw earth from space...

We have been critical of Western technological societies in this book, which means that we, as authors, have created a koan. We happen to live on different continents, and without the inventive power of an industrial society, we would not have the computers, faxes, Internet, and transatlantic aircraft we used to write this book. Revolving now within this koan, having availed ourselves of the technology, our criticism remains. If we have been hard on our own times it is because we have recoiled from the arrogance it often displays, its great desire to control human nature and the material world, its lust for constant progress, and its condescending attitude toward civilizations it classifies as primitive, underdeveloped, or backward. Most of all, we are concerned with our society's obsession with the known and its woeful neglect of the dimension of mystery. It is quite right that we should be amazed at our own achievements, our triumphant technology, our science, the power of our computers, but what we don't know may be sufficiently powerful to overturn in a moment our entire existence and certified knowledge.

Just one quick example of this: At the start of this century, physicists speculated that their subject was coming to an end. Soon there would be no further significant physics to discover. At the time, it seemed there were only a few insignificant pieces of missing information to be filled in: the reason why the orbit of Mercury is irregular, a discrepancy between theory and the amount of heat radiated by a dark object, and the effect of a third body on the motion of two others. Filling in the first piece of missing information produced the theory of relativity; the second, quantum theory; the third led to chaos theory. In each case, the missing information ended up revealing that nature was far subtler and stronger than we had imagined.

Physicists at the end of the nineteenth century had forgotten about hubris. They thought their knowledge could be complete. Meanwhile, the last bothersome bits of missing information they sought remade their world and pumped all the more bits of missing information into it. What "missing information" lies hidden at the end of our own century?

We often joke that someone has a "blind spot," without necessarily realizing that each one of us literally possesses a blind place, an absolute gap in our information-gathering abilities, in the retina of the eye. The retina is packed with visual receptors, except for one tiny region where the nerve connections from those receptors gather together and form the optic nerve. When we look out at the world, there is always a tiny piece of missing information that the brain is constantly filling in so that vision appears to be uniform.

Think of missing information as the trickster of chaos theory. We imagine that we've got everything tied up and accounted for and then the trickster appears to turn things upside down and leap across all our nice convenient boundaries. Like the clown in a medieval court, the trickster is always at our elbow to remind us of our limitations. It is probably not a coincidence—or maybe it's a fractal coincidence—that in the ancient Tarot deck, the card for the trickster, the Fool, the figure with the cap and bells, is also the Tarot's emblem for the whole. The Fool is foolishness or madness, but also spirit. He is the perfected spirit of man approaching the One, the zero that contains all things but is no-thing—the mysterious undefinable chaos.[2]

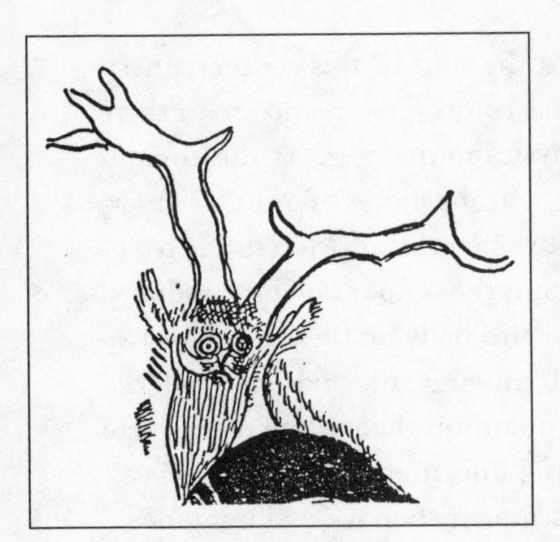

Missing information reminds us that our great achievements are always limited and that one of the most favorable hexagrams in the *I Ching*, the book of changes, is hexagram 15, Modesty.

Modesty is hard to come by in

Photo by John Briggs

our Western civilization, where we take pride in completeness. We want complete scientific theories. Our stories always have to have an ending; music must move toward its final resolutions; until recently, paintings were always bounded within frames. In the Arab world, by contrast, music, art, and storytelling flow on forever without the need for an attainable end point. Although the Sherpas of Tibet like to climb mountains, out of respect to the gods they usually refrain from standing on the peak itself. Imagine a Western climber who didn't have his photograph taken with his foot dominating the summit. Such a person would be considered mediocre and his journey incomplete.

Photo by John Briggs

Perhaps at a deep level our reluctance to embrace the missing information has something to do with our anxiety about death—the ultimate missing information, the ultimate unknown that makes all knowledge shrink to nothing. But chaos theory provides a different outlook. Chaos tells us that the missing information is the window to the whole. In the pit of uncertainty looms our access to creative possibilities.

Everything we have said about missing information and the provisional nature of the concepts we project onto the world obviously must also apply to the metaphor that this book has made of chaos theory. Both the metaphor and the theory from which it came are ways of seeing. A theory is a mental projection onto the infinite complexity of nature—one that emphasizes certain nuances within the flux. The physicist David Bohm liked to point out that the words "theory" and "theater" came from the same

Greek root, "to see." For Bohm, a scientific or a philosophical theory was a "theater of the mind."

A theory is of necessity provisional. It's an abstraction from a much wider context that includes nature, society, and individual life. The context from which theories are born is itself always changing. So theories work well for a time, then seem to get stuck, no matter how much we attempt to modify them, until creative chaos—or whatever else we may call it—causes the mind to come up with a new theatrical production.

But no matter how provocative or insightful, metaphors, theories, concepts, and knowledge can take us only so far. To live sanely and deeply we need something else, a special sort of awareness. Yet as soon as we sense the lack, we immediately ask, "Where does this something else come from? How am I going to grasp it, own it, make use of it in my life." And so the circuit begins again. We jump too quickly from the openness of the question to the need for its resolution. But what if what we are seeking doesn't lie in any answer but at the center of the question, in the very depths of the missing information? Rather than ending this book with a summing up, some definitive statement about life and chaos theory, perhaps we should be simply asking a question.

What question should we ask?

WARNING!!! DO NOT READ THESE WORDS.

Notes

Before Words

1. Quinn, Daniel. *Ishmael.* New York: Bantam, 1992, p. 80.

2. See *Hesiod, Theogony*. Translated by Norman O. Brown. Indianapolis: Bobbs-Merrill, 1953, p. 56.

3. Campbell, Joseph. *Primitive Mythology: The Masks of God.* New York: Penguin, 1987, pp. 273–81.

Lesson 1

1. Krishnamurti, Jiddu. "Krishnamurti Foundation Bulletin," November 1989.

2. Conrad, Joseph. *Typhoon and Other Tales.* New York: New American Library, 1925, p. 22.

3. Conrad's idea that truth about the whole of life is contained in each part is echoed in chaos theory. As we'll see in Lesson 5, chaos tells us that the small scale of things reflects—is self-similar—to the large scale. By looking "sincerely" at the part, Conrad says, we can glimpse the movement of the whole.

4. Shiff, Richard. "Cézanne's physicality: The politics of touch." *The Language of Art History*. Edited by Salim Kemal and Ivan Gaskell. Cambridge: Cambridge University Press, 1991.

5. Medina, Joyce. *Cézanne and Modernism: The Poetics of Painting*. Albany: State University of New York Press, 1995.

6. Merleau-Ponti, Maurice. "Cézanne's doubt." *Sense and Nonsense.* Translated by H. L. Dreyfus and P. A. Dreyfus. Evanston, Ill.: Northwestern University Press, 1964.

7. Fromm, Erich. "The Creative Attitude." *Creativity and Its Cultivation.* New York: Harper & Row, 1959, pp. 54–56.

8. Lao Tzu. *The Way of Life*. Translated by Witter Bynner. New York: Capricorn, 1962, p. 25.

9. Williams, L. Pearce. *Michael Faraday: A Biography*. New York: Basic Books, p. 63.

10. Weber, Bruce. "The Myth Maker." *New York Times Magazine*, October 20, 1985, p. 75.

11. Whyte, David. *The Heart Aroused.* New York: Doubleday, 1990, p. 235.

12. For a complete description of Darwin and the evolutionary tree, see Howard Gruber, "Darwin's 'Tree of Life.'" In *Aesthetics in Science*, edited by Judith Wechster. Cambridge, Mass.: MIT Press, 1978.

13. Csikszentmihalyi, Mihaly. *Creativity*. New York: HarperCollins, 1996, p. 211.

14. Wilhelm, Richard, and Cary F. Baynes, trans. *The I Ching or Book of Changes*. Princeton, N.J.: Princeton University Press, 1967, p. 3.

15. Rubick, Beverly. Personal communication to F. David Peat.

16. Humphrey, N. K. "The Social Function of Intellect." In *Growing Points in Ethology*, edited by P. G. Bateson and R. A. Hinde. Cambridge, Eng.: Cambridge University Press, 1976, p. 312.

17. Shainberg, David. *The Transforming Self*. New York: Intercontinental Medical Books, 1973.

18. Certainly some of the people we recognize as great creators suffered psychological problems in their daily lives, where they were rigid, closed, and self-absorbed. But it seems fair to say that when they were creating, they were open, healthy, and sane. It was through their creative work that they acknowledged their uniquely individual, but entirely indivisible, connection to the whole.

19. Krishnamurti, Jiddu. From September 26, 1948, Poona, India. Provided by the Archives of the Krishnamurti Foundation of America, Ojai, California.

20. Here's a related story. In his book *Games Zen Masters Play*, (New York: New American Library, 1976, p. 123), R. H. Blyth writes: "A monk said to Joshu, 'I have just entered this monastery. I beg you to teach me.' Joshu asked, 'Have you eaten your rice-gruel?' 'I have,' replied the monk. 'Then,' said Joshu, 'go and wash your bowl.' The monk was enlightened." Blyth goes on to point out that the koan looks simple, but can't be resolved intellectually.

"Zen means doing ordinary things willingly and cheerfully. Zen is common life and uncommon life, sense and transcendence, both as one, yet two . . . [Joshu] meant,

> Washing is truth, truth washing; that is all
> Ye know on earth, and all ye need to know.
> The great danger is to divide the washing and the truth."

Lesson 2

1. Lorenz, Edward. *The Essence of Chaos.* London: University College London Press, 1993, p. 14.

2. Harris, Marvin. *Our Kind: Who We Are, Where We Came From, Where We Are Going.* New York: Harper & Row, 1989, p. 44.

3. The quotations of Bolton and Watt decorate the great hall of the Science Museum in Kensington, London.

4. Havel, Vaclav. "The Power of the Powerless." *Open Letters.* New York: Vintage, 1992, p. 132.

5. Whyte, David. *The Heart Aroused.* New York: Doubleday, 1990.

6. Patterson, Michael. Personal communication with John Briggs.

7. Another type of negative influence is felt when a person responds to the mob consciousness and madness of crowds. That person has given up his individuality and identified totally with the group. The mob is a limit-cycle system.

8. Havel, p. 147.

9. Meeker, Joseph W. "The Comedy of Survival." *Search of an Environmental Ethic.* Los Angeles: Guild of Tutors Press, 1980.

10. Parks, Rosa. *Quiet Strength.* Grand Rapids, Mich.: Zondervan Publishing House, 1994.

11. Lewis, Anthony, and the *New York Times. Portrait of a Decade: The Second American Revolution.* New York: Bantam Books, 1965, p. 63.

12. Musil, Robert. *The Man Without Qualities*, Vol. 1. Translated by Burton Pike. New York:, Vintage, 1995, p. 7.

13. Compare Parks's action with the terrorist who blows up a building in the hope that his action will set off an uprising. The terrorist is certainly not "living in truth," but in the fantasy that he can turn his powerlessness into power and exercise control over the situation through a violent act.

LESSON 3

1. Pelletier, Wilfred, and Ted Poole. *No Foreign Land.* New York: Pantheon, 1973, p. 199.

2. Woodman, Lynda A. "Business and Complexity," in doctoral documents submitted November 1996, The Union Institute.

3. Brian Goodwin is quoted in Roger Lewin, *Complexity: Life at the Edge of Chaos.* New York: Macmillan, 1992, p. 41.

4. Angier, Natalie. "Status Isn't Everything, at Least for Monkeys." *New York Times*, April 18, 1995, p. C6.

5. Here's an interesting illustration of the fact that competition is in the eye of the beholder: To Americans, the traditional Western film plot seems to center on a competition to the death between heroes and villains. Sociologists point out that a Japanese audience may come away from the same film with a moral about the virtues of cooperation.

6. We should note that, in reality, even mechanical systems such as pistons aren't as regular as they appear in scientists' graphs found in high school science text books. Scientists like to deal with idealized systems in which pistons work without friction and pendulums don't meet air resistance. Real mechanical systems, however, are subject to the contingencies of the world and the individuality of their particular construction. In practice, their behavior, too, shows evidence that a strange attractor lurks behind the scenes.

7. Mander, Jerry. *In the Absence of the Sacred.* San Francisco: Sierra Club, 1992. In an essay in a newsletter *Touch the Future* (Long Beach, Calif., Fall 1997), the internationally known children's entertainer Raffi describes the effect of TV in terms of a limit cycle: "Our high-fat media diet creates a 'virtual' reality, a giant negative feedback loop expanding and feeding on itself. Life as performance, relationship as marketing, a techno-babble melodrama where time and complexity are compressed into a numbing tedium of pseudo-crises and shopping solutions." One wonders whether TV, with its hyped-up competitions, hasn't led us to devise a free-market version of Havel's post-totalitarian society.

8. Quoted in J. P. Mayer, *Max Weber and German Politics.* London: Faber and Faber, 1943, pp. 127–28.

9. Whyte, David. *The Heart Aroused.* New York: Doubleday, 1990, p. 21.

10. Ibid., p. 262.

11. Ibid., p. 296.

12. Bower, Bruce. "Yours, Mine and Ours." *Science News*, Vol. 153, No. 3 (March 28, 1998), pp. 205–207. A new generation of economists and social scientists are discovering that people are in fact naturally inclined to cooperate for the common good. These findings challenge the Darwinian assumption that economic behavior is only "rational" if it is selfish and self-interested.

In one study, researchers invented economic games where participants were kept unaware of each other's choices to cooperating on a common task or acting selfishly, as traditional economic theory says they will naturally do. When volunteers received money in private accounts, most instinctively chose to contribute to a public fund, giving themselves to the chaos of a collective uncertainty. The consequence was that they all received modest financial rewards. Selfish players who didn't donate received greater personal advantage, but they ended up disrupting the community. As the game progressed, other players diminished their own public contributions in a futile attempt to punish the exploiters. This caused the previously unselfish players to end up with less money than if they had continued to pay into the public fund.

Traditional economic theory assumes that indigenous hunters who share large parts of their kills with comrades must only do so in the selfish expectation of future preferential treatment. But evolutionary biologist David Sloan Wilson says it is more reasonable to assume that the hunters share to help their group, not themselves. Perhaps our rationality is more generous than economists would have us believe.

13. Wheatley, Margaret, and Myron Kellner-Rogers. *A Simpler Way*. San Francisco: Berrett-Koehler, 1996, p. 57.

14. Whyte, p. 243.

15. David Bohm interviewed by John Briggs. "Dialogue as a Path to Wholeness." *Discovering Common Ground.* Edited by Marvin R. Weisbord. San Francisco: Berrett-Koehler, 1992.

16. Shainberg, David. Personal communication with John Briggs.

17. "Collaborative memory" provides another kind of example of collective self-organization. Researchers have found that elderly spouses compensate for the normal decline in the ability to recall familiar names, faces, and events. They do this by becoming experts at using feedback to collaborate with each other so they can reconstruct what's missing. In one study, couples married for between thirty and fifty years were able to remember much more of a short story read to them than young individuals. "The recent findings challenge current scientific assumptions about the mind," contends psychologist Laura L. Carstensen of Stanford University. Instead of treating mental activity solely as the product of individual brains, she remarks, investigators should explore whether the mind exists first in social interactions that influence what individuals think and do." (Bruce Bower, "Partners in Recall," *Science News*, Vol. 152, No. 11 [September 13, 1997], pp. 174–75.)

Lesson 4

1. The anecdote about Frochlich was related to David Peat by his friend and colleague Thomas Grimley.

2. Scientists studying flour beetles found that chaos explains why it is difficult to eliminate crop pests. When researchers applied pesticide to adult flour beetles, the population didn't decrease but fluctuated wildly. The more adults they killed off, the bigger the fluctuations.

3. Heron, Patrick. "Solid Space in Cézanne." *Modern Painters*, Vol. 9, No. 1 (Spring 1996), pp. 16–24.

4. Compare the Buddha's statement with eighteenth-century philosopher David Hume, who wrote that the self is "nothing but a system or train of different perceptions," a fantasy, a fiction of the imagination rather than an ineffable category all its own (1739). Echoing Hume, Daniel Dennett, a modern theorist of consciousness, calls the self "an abstraction" or a "Center of Narrative Gravity." (See Galen Strawson, "The Self," *Journal of Consciousness Studies,* Vol. 4, No. 5–6 [1997], pp. 405–28.)

In contrast to these approaches, chaos suggests that the self *does* exist and *is* real but not as a fixed entity—rather as a movement of interconnection fluctuating somewhere between the sensations of the solitary, unique experiences and the inflowing of the social human consciousness that we all share.

5. Lopez, Barry. *Arctic Dreams: Imagination and Desire in a Northern Landscape.* New York: Bantam, 1996, p. 181.

6. Shah, Idries. *The Way of the Sufi.* New York: E. P. Dutton,1970, p. 122.

Lesson 5

1. Sirén, Osvald. *The Chinese on the Art of Painting.* New York: Schocken, 1963, p. 2.

2. An actual piece of paper, of course, has three dimensions. Its depth dimension is very small.

3. Conrad, Joseph. *Typhoon and Other Tales.* New York: New American Library, 1925, p. 20.

4. Shepherd, Linda Jean. *Lifting the Veil: The Feminine Face of Science.* Boston: Shambhala, 1993.

5. DNA is sometimes cited by complexity scientists as an example of how a simple rule or algorithm iterated with variations generates a multiplicity of organic forms. At bottom, this argument is meant to suggest that nature is essentially a collection of sophisticated chemical algorithms that we can mimic with our sophisticated mathematical algorithms. The problem here is that we've gotten used to thinking of DNA in a very simplified way. As discussed in Lesson 4, oversimplification leads to distortion. The DNA molecule is in feedback relationship with countless forces and processes working to create a living form. A subtler way to think of DNA is as one of the multitude of fractal microcosms reflecting the individual.

6. We've used the expression "poetic and artistic metaphors" to refer to metaphors that have an active tension between similarities and differences. Artistic metaphors are multilayered, self-contradictory from a logical point of view, vivid, and affecting. But there are other kinds of metaphor. For example, everyday metaphors that colorfully describe things ("He's as alive as a cricket") or metaphors that join ideas or images in a provocative way in order to make a point (comparing chaos theory to Zen or poetry, or writing a book using chaos theory as a metaphor). In nonpoetic metaphors, the similarities between the terms are ultimately the "point" of the metaphor. These types of metaphors are a colorful way of making new categories or illustrating new abstractions. Poetic metaphors, in contrast, subtly subvert categories and abstractions in order to get beyond them. Natural fractals have self-similarity at different scales. But the self-similarity of art isn't the same as the kind of scaling found in a tree. We could think of art as having, rather, many different "scales of abstraction." Consider, for example, T. S. Eliot's image from "The Love Song of J. Alfred Prufrock." Eliot describes the self-

conscious Prufrock mortified by the power of others to judge him: "The eyes that fix you in a formulated phrase,/And when I am formulated, sprawling on a pin,/When I am pinned and wriggling on the wall." These lines contain three metaphors. We can follow them as a nest of self-similarities: Eyes are like a formulated phrase, eyes and the phrase are like a pin, the speaker is like an insect. Each term in these sets of metaphors (eyes, phrase, formulated, insect) represents a different type or "scale" of the many levels of categories we use to describe the world. The self-similarities and dissimilarities (the metaphors) are made by illogically combining items in these levels of category. So a metaphor creates something *like* the self-similarity we find at different scales in the natural environment, but the self-similarities made by metaphors also have their own unique quality.

7. Beethoven undoubtedly could create an enduring sense of spontaneity in his concertos and quartets, because he himself experienced such spontaneity, even after going through the piece numerous times in the act of composition. Many artists have said this. The piece continues to surprise them as much as it surprises the audience. They have made the piece by subverting their own algorithms, their own abstractions. But how that happens is one of the great mysteries (or perhaps trickster secrets) of art. The creative self-similarity and difference at work within the fugue and between the fugue and the listener is also at work in a great Indian raga, African drumming, symphonies by Romantic composers like Brahms or Beethoven, or in a modern twelve-tone composition. Arnold Schoenberg, who introduced the twelve-tone system to history, said that in music "dissonances are the remote of consonances" and "whatever happens in a piece of music is nothing but the endless reshaping of the basic shape." Schoenberg echos Virginia Woolf's assertion that in her writing "I attain symmetry by means of infinite discords, showing all traces of the mind's passage through the world; achieve in the end some kind of whole made of shivering fragments." Woolf stresses the sense of

life or "being" that permeates a work of art, as if it were a work of nature (which, of course, it is).

8. Despite the power of Newtonian reasoning, threads of such an aesthetic rationality have woven their way through the last two hundred years. Wolfgang Goethe, for example, was highly critical of Newtonian science, arguing that instead of allowing the natural world to speak to us directly, science gained its knowledge through experiments that forced nature into artificial contexts. By means of these experiments and detailed observation, science seeks an abstract unity that it believes must lie behind the diversity of nature. Goethe, in contrast, argued that it is possible to develop a sense of empathetic unity with nature, an aesthetic sense, that allows us to perceive unity directly and vividly, as an actuality rather than an abstraction. The contemporary biologist Brian Goodwin agrees, arguing that an "objective intuition," something along the lines proposed by Goethe, could be used to supplement orthodox biology's methods of analysis. A plant can be approached both in its analytical abstractions and in an overall aesthetic way that gives attention to what could perhaps be called the plant's "meaning" or "significance" in relationship to the natural world.

9. The scientists who developed the computer model didn't believe that high-altitude hydrocarbon spraying was a practical solution, but they did hope that it would start a serious discussion leading to a technological way of solving the ozone-hole problem. This would certainly decrease the sense of urgency about reducing fluorocarbon emissions. ("Refilling the Ozone Hole," *New York Times*, November 26, 1991, C2–3.)

10. Alexander, Christopher. *The Timeless Way of Building.* New York: Oxford University Press, 1979, p. 135.

11. Ibid., p. 137.

12. The importance of art as a way to gain insight into the natural order has been recognized by many scientists, including Mandelbrot and Belgian scientist Ilya Prigogine, the Nobel laureate who introduced many of the key ideas of chaos in the 1970s.

13. We add here a fanciful speculation. What if enlightenment is actually the experience of seeing in one glance that all of creation is a kind of fractal and realizing that at every scale it is both different and the same.

Lesson 6

1. Creative artists often describe the way the entire vision of a piece comes to them all at once, even though its details must later be unfolded in time. Mozart claimed that entire symphonies and concertos came into his head and he simply remembered and transcribed them. The American musician Therese Schroeder-Sheker describes how a piece of music arrived to her complete, yet "it was lying outside time." The music was present, but not in the usual sense of a temporal sequence of sounds. Later, the piece would have to be written down within the linearity of time.

The British composer Sir Michael Tippet referred to his experiences while composing as "possession." "There is no invocation, no act of will. [T. S.] Eliot and I talked about it a lot. He said, 'The words come last.' On the contrary, with me it just appears—like that and I must accept it." These accounts come from the creators' interviews with David Peat.

2. Erich Fromm tells the story of a woman who dreamed a monster was sitting at the foot of her bed. "Help! Help!" she

cried. "What are you going to do with me?" The monster shrugged, "It's your dream, lady." Psychotherapists have often argued that within a dream, each character and element represents the dreamer. In their book *Individual Object Relations Therapy* (forthcoming from Jason Aronson), psychiatrists Jill and David Scharff propose that each dream is a fractal manifestation of the dreamer's entire personality.

3. Proust, Marcel. *Swann's Way*. Translated by C. K. Moncrieff. New York: Modern Library, 1956, p. 65.

4. Shenk, David. "Life at Hyper-Speed." *New York Times*, September 19, 1997, p. A35.

5. Russel, Andy, in conversation with David Peat. According to an often repeated story, a visitor who watched Picasso make a rapid drawing asked him how much it would sell for. When Picasso named a sum in the hundreds of thousands of dollars, the visitor reacted, "Fancy making so much money in just two minutes." "But," Picasso answered, "it took me sixty years of hard work to make that drawing."

Lesson 7

1. Kelley, Kevin W., ed. *The Home Planet.* Reading, Mass.: Addison Wesley, 1988, p. 138.

2. Ibid., p. 26.

3. Capra, Fritjof. *The Web of Life.* New York: Anchor, 1996, p. 4.

4. Panofsky, Erwin. *Abbot Sugar on the Abbey Church of St. Denis and Its Art Treasures.* Princeton, N.J.: Princeton University Press, 1946.

5. The evolution of the idea of the self is brilliantly outlined in a paper by Roy F. Baumeister published in the *Journal of Personality and Social Psychology* (Vol. 52, No. 1 [1987], pp. 163–76). In the medieval period, the self was "unproblematic," according to Baumeister. It was equated solely with the public self—a self engaged in demonstrating that it possessed the morality and virtue that would lead to Christian salvation. In the early modern period (sixteenth to eighteenth centuries), people began noting a difference between their true inner selves and their outer apparent selves. This led to a societal craving for "sincerity" as a way of joining these two selves. Around this time, people became interested in creativity as a means of fulfilling the inner self. During this period, our modern sense of self began emerging. Baumeister remarks that "an abstract, hidden self is harder to know and define than is a concrete, observable self. Therefore, the belief in a real self that is hidden, that is not directly or clearly contained in one's action, can be regarded as a critical complication of self-knowledge. The inner nature of selfhood, which is regarded as axiomatic by much modern psychological thought, seems to have become a common conception first in the sixteenth century." Gradually, privacy became an issue for individuals because "until the end of the seventeenth century, nobody was ever left alone." Buildings began to be constructed with privacy in mind.

With the Puritans, the self turned inward, into self-consciousness and a concern with self-deception. Self-knowledge became increasingly uncertain. During the Romantic period in the late eighteenth and early nineteenth centuries, many sought fulfillment of the self as a lifetime goal. Creativity or the passion of romantic love became the chief means of this fulfillment as the self set out to discover its own destiny. The individual was now posed against society, struggling for freedom. This was the period of the American Revolution. Says Baumeister, "That 'all men are created equal,' a notion labeled a self-evident truth' in the revolutionary manifesto of the American colonists (1776), would have been inconceivable to the medieval mentality."

During the Victorian period of the mid- and late nineteenth century, the self became viewed as hypocritical and individuals tried to transcend it.

In the twentieth century, Freud's theories fostered a belief in the literal impossibility of self-knowledge. The self was increasingly seen as isolated and alienated from society.

In the present era, according to Baumeister, the idea of the self has undergone still further evolution. Our belief in personal uniqueness has been intensified and we now seek fulfillment of the self in celebrity and "self-actualization." We define ourselves in terms of our personalities and our socioeconomic status or accomplishments.

In a 1977 book entitled *Evil, Inside Human Violence and Cruelty* (New York: W. H. Freeman), Baumeister adds a further disturbing twist to the idea of the self as a construction. He proposes that evil is done by individuals who have both an inflated idea of self, or egotism, and a heightened sensitivity to perceived slights. Such people easily rationalize their behavior, interpreting its cruelty as something done for the "supremely good." Baumeister's idea of an overinflated view of self and the potential for evil would apparently apply to a public medieval self, a "sincere" sixteenth-century self, or a modern alienated one, though the evil would have a different character and meaning in each case.

Of course, Baumeister's history of the Western self can't show us—the way a Shakespeare play does—the immense subtlety of the subject. For example, it can't show us the many shades of ideas of self that are current at any one time in a culture. But it does help dramatize the degree to which some of our most intimate and urgent senses of self are historical and social constructions. What new sense of self might chaos help construct?

6. Eco, Umberto. *Art and Beauty in the Middle Ages.* New Haven: Yale University Press, 1986.

7. Blakeslee, Sandra. "Some Biologists Ask 'Are Genes Everything?'" *New York Times*, September 2, 1997, C1.

8. For discussions of Gaia see: James Lovelock, *Gaia: A New Look at Life on Earth* (Oxford: Oxford University Press, 1979) and *The Ages of Gaia: A Biography of Our Living Earth* (New York: Bantam, 1988), and Lynn Margulis and Dorion Sagan, *Slanted Truths: Essays on Gaia, Symbosis and Evolution* (New York: Springer-Verlag, 1997).

9. Margulis, Lynn, and Dorion Sagan. *Microcosmos.* New York: Summit, 1986.

Chaos theory makes an important distinction between the "structure" of an organism, or system, and its "organization." This distinction helps to clarify the difference between the new holism of chaos theory and the mechanistic view the West has been immersed in for so long.

The "organization" of a living system is not so much its particular components (tissues and organs, for example) but the system of feedback relationships between them. A factory, an airline, a film crew, and a living cell all look totally different, yet in terms of the feedback links within each system and the dynamical flows of material and information, they may be strikingly similar. Whereas the old mechanistic view placed its focus upon physical components and their mechanical interactions, the new vision concentrates on dynamical processes, movement, and flow.

10. We're not referring here to the natural differences in point of view that are involved in any dispute. Chaos theory suggests these unique points of view are invaluable, as we discussed in Lessons 1, 2, and 3. From the chaos perspective, individual differences actually form the basis for the resolution. In the old mechanical context, however, differences can only be resolved through competition that leads to conquest or compromise.

11. Gore, Rick. "Andrew's Aftermath." *National Geographic*, April 1993, p. 25.

12. Just as feedback can unify people, stimulating them to help others, it can also unify mobs and toxic political groups like the Nazis. The diverse individuals drawn to aid the victims of Hurricane Andrew are expressing a deep sense of unity that (momentarily, at least) surpasses divisions such as race, class, and sexual orientation. The feedback they exhibit is part of a vast "open system." The Nazi vision of the whole was without diversity and freedom. Here, feedback linked individuals in a powerful limit cycle.

13. Frankl, Victor E. *Man's Search for Meaning*. Translated by Ilse Lasch. New York: Washington Square Press, 1963, p. 59.

14. Conrad, p. 20.

After Words

1. Jung, Carl. *Analytical Psychology, Its Theory and Practice: The Tavistock Lectures*. New York: Vintage, 1970, p. x.

2. Cavandish, Richard. *The Black Arts*. New York: Capricorn, 1968, p. 114.

INDEX